RUNNING, WALKING AND JUMPING

The Wykeham Science Series

General Editors:

PROFESSOR SIR NEVILL MOTT, F.R.S.
Emeritus Cavendish Professor of Physics
University of Cambridge

G. R. NOAKES
Formerly Senior Physics Master
Uppingham School

Biology Editor:

W. B. YAPP
Formerly Senior Lecturer
University of Birmingham

The Author

ANNE INNIS DAGG became interested in animal locomotion while studying the giraffe in Africa. She is now Research Assistant Professor at the University of Waterloo, in Ontario, Canada.

The Schoolmistress

ANNE C. JAMES is a biologist at Queen Anne Grammar School, York.

RUNNING, WALKING AND JUMPING

The Science of Locomotion

ANNE INNIS DAGG

University of Waterloo, Canada

WYKEHAM PUBLICATIONS (LONDON) LTD
(A member of the Taylor & Francis Group)
LONDON and BASINGSTOKE   1977

Distribution and Representation

UNITED KINGDOM, EUROPE AND AFRICA
Chapman & Hall Ltd. (a member of Associated Book Publishers Ltd.),
11 North Way, Andover, Hampshire SP10 5BE.

UNITED STATES OF AMERICA AND CANADA
Crane, Russak & Company, Inc.
347 Madison Avenue, New York, N.Y. 10017, U.S.A.

AUSTRALIA, NEW ZEALAND AND FAR EAST
Australia and New Zealand Book Co. Pty. Ltd.,
P.O. Box 459, Brookvale, N.S.W. 2100.

JAPAN
Kinokuniya Book-Store Co. Ltd.,
17–7 Shinjuku 3 Chome, Shinjuku-ku, Tokyo 160–91, Japan.

INDIA, BANGLADESH, SRI LANKA AND BURMA
Arnold-Heinemann Publishers (India) Pvt. Ltd.,
AB-9, First Floor, Safdarjang Enclave, New Delhi 110016.

GREECE, TURKEY, THE MIDDLE EAST (EXCLUDING ISRAEL)
Anthony Rudkin, The Old School, Speen, Aylesbury,
Buckinghamshire HP17 0SL.

ALL OTHER TERRITORIES
Taylor & Francis Ltd., 10–14 Macklin Street, London WC2B 5NF.

Contents

		Page
Preface		ix

Chapter 1 HISTORICAL APPROACH

1.1	Prehistoric Man	1
1.2	Early Civilization	2
1.3	Renaissance Europe	3
1.4	Basic Research	4
1.5	Current Research	6
1.6	Current Concepts	8

Chapter 2 ADAPTATIONS OF LOCOMOTION

2.1	Use of Locomotion	10
2.2	Maximum Speeds	10
2.3	Is Fast Best?	11
2.4	Is Locomotion Limiting?	13
2.5	Endurance	13
2.6	Other Adaptations	14

Chapter 3 MORPHOLOGY OF VERTEBRATES

3.1	Basic Materials for Locomotion	15
3.2	Evolution for Improved Locomotion	16
3.3	Bones and Locomotion	19
3.4	Muscles and Locomotion	23
3.5	Action of Muscles on Bones	24
3.6	Nervous Control of Locomotion	25

Chapter 4 HUMAN WALKING AND RUNNING

4.1	Bipedal Locomotion	27
4.2	Primate Evolution	27
4.3	Methods of Studying Human Gait	29
4.4	Centre of Gravity	31
4.5	Walk—Definitions	33

Page

4.6	Walk—Gross Displacements	34
4.7	Walk—Pelvic Movements	35
4.8	Walk—Force and Energy Changes	36
4.9	Walk—Muscle Action	39
4.10	Walk—Variations	40
4.11	Walk—Medical Research	41
4.12	Running	42
4.13	Locomotion on the Moon	45

Chapter 5   ATHLETICS

5.1	Sprinting	48
5.2	Long-distance Running	50
5.3	Hurdling	51
5.4	Long Jump	52
5.5	High Jump	54
5.6	Racial Differences	55

Chapter 6   WALKING AND RUNNING IN FOUR-LEGGED MAMMALS

6.1	General Principles	56
6.2	Gaits and Gait Analysis	59
6.3	Walk	63
6.4	Trot	69
6.5	Pace	70
6.6	Gallop and Bound	71
6.7	Choice of Fast Gait	75
6.8	Energy Relationships	75
6.9	Occasional Bipedalism	77
6.10	Display Gaits	77
6.11	Neck Movements	78

Chapter 7   WALKING AND RUNNING IN LOWER VERTEBRATES

7.1	Evolution and General Characteristics	81
7.2	Amphibians with Tails	81
7.3	Frogs and Toads	83
7.4	Lizards	84
7.5	Crocodiles	86
7.6	Tortoises and Turtles	87
7.7	Dinosaurs	89
7.8	Birds	91

Page

Chapter 8 **HOPPING AND JUMPING IN VERTEBRATES**

8.1	Filming in Zoos	98
8.2	Jumping, Hopping, and Leaping	98
8.3	Ricochet	99
8.4	Analysis of Kangaroo Jump	101
8.5	Kangaroo Morphology and Evolution	102
8.6	Jumping Small Mammals	104
8.7	The Jump of Frogs	107
8.8	Jumping over Obstacles	111
8.9	Jumping on to an Object	111

Chapter 9 **TRACKS OF VERTEBRATES**

9.1	Identification of Tracks	113
9.2	Classification of Gaits	115
9.3	Records of Tracks	115
9.4	Research Involving Tracks	118

Chapter 10 **INVERTEBRATE LOCOMOTION**

10.1	Onychophoran Gaits	120
10.2	Primitive Arthropods	122
10.3	Insect Anatomy	122
10.4	Insect Locomotion	125
10.5	Locomotion of the Ghost Crab	127
10.6	Locomotion of a Crayfish	129
10.7	Locomotion of a Spider	130
10.8	Jumping in Invertebrates	130

FURTHER READING	135

Preface

LOCOMOTION is one of those subjects which touch on various established disciplines, but belong completely to none. Workers on one facet of movement studies may be completely unaware that other facets are also being studied. Researchers in pathological gaits usually know nothing about zoological studies on the movements of wild animals; coaches in athletics may be oblivious of anatomical studies on muscles and joints; men who breed and race thoroughbreds often have no idea that there is scientific information on horse movements that could help them. The aim of this book is to deal with the many varied approaches to the study of locomotion so that workers in the field can be aware of alternative ways to approach the subject. Students will realize the breadth of work on locomotion, and perhaps be able in the future to fill in some of the many gaps in our knowledge. There is a great deal to be done.

The names given to the animals mentioned in a text such as this are always a problem. How explicit should one be, without being pedantic? If a particular species of animal is referred to here, the scientific as well as the common name is given at least once. If common animals in general are discussed (e.g. elephants, horse, dog, etc.) only the common name is used. All of the names, both scientific and common, are listed in the index. To avoid confusion among North American and European readers, the word elk is not used; moose refers to *Alces alces* (= elk in Europe) and wapiti refers to *Cervus canadensis* (= elk in North America).

There are numerous books dealing with human movement and athletics, many of them available at good public or university libraries. Books on running, walking, and jumping in four-legged animals are less well known, so these are listed on page 135. Details on specific research projects mentioned in the text are available either from the reference source mentioned in related figures, or from the author.

I am very grateful to all those research workers who gave me permission to reproduce their figures in this book; their names and the references to their work are given in the various figures. Many other people kindly helped me in my own research on locomotion, particularly Mr. D. Leach, Mr. J. Roden, and Mr. H. Dagg; this work could not have been carried out but for a generous grant from the National Research Council of Canada, aid for which I am deeply indebted. My thanks also go to Ms. Rosemary Allen, who prepared most of the figures, to Ms. Betty Sampson who typed the manuscript twice, to Ms. Jane Schmalz who typed many dozens of letters connected with this project, and to Dr. I. R. Dagg and Miss P. M. Brown who gave advice on the sections dealing with the physics of animal movement.

ANNE INNIS DAGG
1976

1. Historical Approach

1.1. *Prehistoric Man*

Ever since man's ancestors evolved into creatures which hunted animals, they have been concerned with their prey's locomotion. Their method of hunting each species depended upon the movements it made. Sloths for example could readily be killed because they moved so slowly. Kangaroo species that paused in their flight to look back were more vulnerable than those that fled without hesitation. The hares that ran under fallen logs could be snared more easily than those that jumped over them. The game that moved along trails would be likely to fall into pit traps, unlike the game that moved more randomly. Some species could be stampeded over cliffs, while others would turn back on their pursuers. Slow animals or animals with little endurance offered a better chance for human hunters than did fast animals that could run long distances.

Primitive man killed the animals he could catch. Some species were so vulnerable to his weapons and strategies that he apparently wiped them out entirely—species such as cave bears from Eurasia; giant tortoises, ground sloths, camelids, and mammoths from North America; moas from New Zealand; and giant marsupials from Australia.

Once their skill at hunting was great enough to allow them some leisure, about thirty thousand years ago, prehistoric men began to draw pictures of animals on rocks or cave walls. No one knows why these people did this—perhaps to ensure fertility of a species, or to learn about it, or to assume some supernatural power over it, or to gain strength from it. At first these pictures were simple depictions of species that man encountered in hunting; he did not bother with small mammals or with most birds. His hunter's eye was keen, so that often even a few simple lines created an animal which was amazingly life-like. The artist who drew the ostriches which once inhabited the Sahara captured superbly the way in which they ran and held their bodies (fig. 1.1). It is easy by studying the pictures to note a shift of culture in the Sahara from an earlier hunting society when giraffe, antelope and rhinoceros were drawn, to a later pastoral one when sheep and cattle were depicted.

Fig. 1.1. Ostrich running. Drawn on rocks in the Sahara by prehistoric man. Their motions are fluid, with their legs bending as they do in real life.

The urge to draw the animals with which primitive man was familiar must have been great, because rock pictures have been found throughout the world—kangaroos in Australia, bison in Europe, giraffe in Africa drawn by bushmen, deer drawn by North American Indians. All the animals were drawn either stylistically or as naturally as possible. The latter were usually depicted standing still; this was easier to do than drawing running animals, and it was the way hunters saw their prey as they laboriously stalked them. When they did illustrate running quadrupeds it is usually impossible to tell if their leg postures are portrayed accurately.

1.2. *Early Civilization*

With the rise of civilizations in the Mediterranean area, man's portrayal of animals became generally more sophisticated and detailed. The Egyptians, like primitive men, drew animals with religious feeling, but they made them stiff and conventional in design. Artists in Mesopotamia drew them more abstractly, often incorporating the parts of many animals into one strange individual. The Minoans and Assyrians depicted animals realistically. The Assyrian horses (fig. 1.2) are well portrayed in themselves with good detail of muscles and tendons, but their legs are not correctly drawn. Both horses are obviously walking, but they are doing so in a pacing manner which would never be used in real life. The animals are having to support themselves on their two right legs while the two left swing forward together. In reality the left hind leg is always swung forward before the left front leg in walking hoofed animals. A pacing gait is possible in horses, but only in ones that are moving much faster than these, held back as they are by the man striding along between them.

Fig. 1.2. Assyrian horses being led by a warrior. From a seventh century bas-relief in the Louvre, Paris. The left legs are both in the air at once, which would make balance at a slow speed almost impossible.

Aristotle was probably the first person to write about locomotion. He felt that because the right side of an animal was naturally superior to the left, all locomotion began from the right. In fact his definition of right was the part from which locomotion began. He believed men tended to carry burdens on their left shoulder and stand on their left legs, so that the more active motor side would remain unencumbered. When they wanted to fight, men advanced their left legs and kept their right hands back, preparatory to a blow.

1.3. *Renaissance Europe*

With the rise of Christianity in the Mediterranean world, artists and scholars lost interest in natural history. Animals in pictures were used largely as symbols; there was little effort at naturalism, so that their movements (except for the flight of birds) were seldom illustrated. Little was written on locomotion although Leonardo da Vinci speculated on human movements. By 1657 however, P. G. Newcastle had worked out a detailed description of gaits in horses, in part by listening to their hoofbeats. Shortly after this Giovanni Borelli wrote *De Motu Animalium* (1679) in an attempt to apply strictly mechanical analyses to the movement of animals, using the principles of screw, lever, pulley, scale and wedge to demonstrate the action of muscles and the mechanics of locomotion.

It should have been possible for early artists to determine how a large quadruped's legs moved in relation to each other when it was walking, because each leg is moving relatively slowly. Such details apparently did not concern them however. Nor did it concern many artists in pre-Renaissance and Renaissance Europe. These men and women were less familiar with animals than their forebears had been and they often worked in studios rather than directly from their subjects. Giorgio Vasari, commenting in the 16th century on Uccello's large painting of Sir John Hawkwood on horseback, wrote, "This work was and still is held to be something very beautiful for a painting of that kind, and if Paolo had not made that horse move its legs on one side only, which naturally horses do not do or they would fall—and this perchance came about because he was not accustomed to ride; nor used to horses as he was to other animals—this work would have been absolutely perfect."

1.4. *Basic Research*

The first serious quantitative work on locomotion in man was published by Wilhelm and Ernst Weber in 1836. These German brothers collected a great deal of data from both cadavers and living people. They introduced among other work the *Pendulum Theory*, which stated that the motion of each leg as it swings forward is a pure pendulum motion not dependent upon muscular action. With the later development of cinematography which enables accurate photographs and measurements to be made of the positions of a swinging leg over short periods of time, this theory was repudiated.

Forty years after the Webers' studies, Professor E. J. Marey of France developed what he called *chronophotography*, which would not only be used to reanalyse the Webers' work on leg movements, but which would be the basis of almost all detailed studies of locomotion from that time to the present. In Marey's invention successive exposures were made on a single photographic plate by means of a disc rotating in front of the camera. During a sunny day a person dressed in black, with shiny metal buttons and bands fastened to his costume at various places to represent joints and specific bones, was photographed while walking in front of a black screen. The resulting photographs formed geometric pictures of lines and points which could be measured to give a great deal of information about the person's movements during successive small units of time.

At this period another father of locomotion studies, Eadweard Muybridge, was using a battery of 24 still cameras ingeniously triggered in sequence to film animals either walking or running. His classic *Animal Locomotion*, published in 1887, contained 781 plates with from 10 to 48 separate photographs on each. Each volume cost £100, so the work was never popular.

Despite Muybridge's new techniques for studying quadrupeds, research in locomotion during the next fifty years was largely devoted to man. Otto Fischer analysed human movements mathematically at the turn of the century and together with W. Braune, an anatomist, produced an important work on human gait. More recently their research has been extended by N. A. Bernshtein and his associates in Moscow, and by Herbert Elftman in United States.

If the slow movements of quadrupeds were difficult to interpret accurately without the help of photography, the fast gaits were impossible to decipher. Indeed, one of the first controversies which photography solved was whether or not a horse while trotting could have all its legs off the ground at the same time. Before this was shown to be possible by Muybridge in 1872, and even after, artists were drawing galloping horses in positions never assumed in nature. Their horses looked swift, with their front legs straining forward and their hind legs stretched out behind, but given the force of gravity it would be difficult to see, for example, how the galloping horse from Gericault's work drawn in fig. 1.3 could but end up sprawled on its stomach. A horse's legs are spread out front and back when it is jumping, but then it is going up rather than along. Only slender animals with flexible spines, such as greyhounds, cheetah, deer and hares, can gallop or bound in the manner 19th century artists drew their racing horses.

It is possible that the idea of a horse going quickly with its legs spread out as pictured in fig. 1.3 is a naïve one, and present in children who have only visualized swift horses in their mind's eye, not studied their movements more analytically. To determine if this were so, I asked a number of children under twelve years of age to draw a horse running as fast as it could. Many children refused even to attempt such a drawing. Of those who were willing to try, only children over six years of age drew recognizable animals. Eighteen of these horses had their legs under them (shown with an amazing variety of bends and curves), while only two had them spread out as early artists depicted them. This artistic posture therefore seems to be one adopted

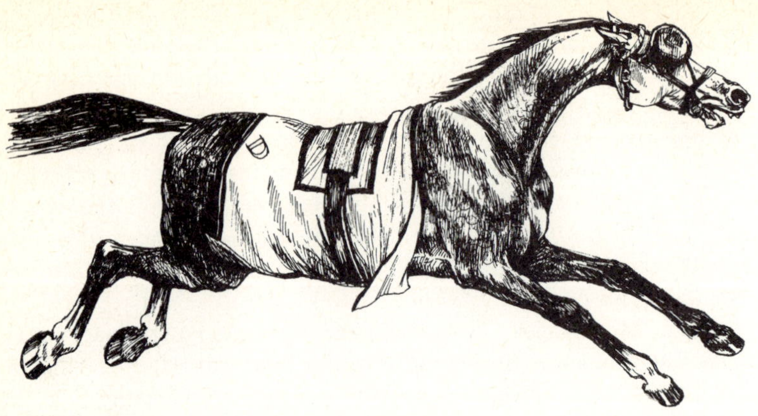

Fig. 1.3. Exercising horse. From an engraving by Gericault. No horse in real life ever gallops with its legs spread out in this manner.

consciously by artists who are more interested in a feeling of swiftness than in accuracy.

1.5. *Current Research*

With increased technology the study of locomotion in man has advanced greatly since World War II. At present anatomists, kinesiologists, biophysicists, and neurophysiologists are all producing important work which affects not only athletes and people unable to walk normally, but even astronauts who have to move under reduced gravity conditions (as on the Moon) and even under zero gravity within satellites. The clinical laboratory for locomotion at the Shriners Hospital for Crippled Children in Winnipeg, Canada, has developed a sophisticated apparatus for analysing the movements of abnormal children (fig. 1.4). While the child moves along a 7-metre runway, television cameras record his progress from the front and the side. The changing position of markers which have been fastened at fixed sites to his legs are relayed to videotape recorders. A biotelemetry system is attached to the child's waist, with wires running to several muscles in his legs and to his shoes, so that signals indicating the time of contraction of each muscle in relation to the phase of a stride can be transmitted to the control centre where they are recorded on an oscilloscope. These techniques will be described more fully in Chapter 4.

In Winnipeg the apparatus has been set up so that clinicians can assess a child's problems quickly. The data can also be converted into printed form so that his gait can be analysed more thoroughly after he has left the laboratory. Thus tracings can be made from the oscilloscope, and

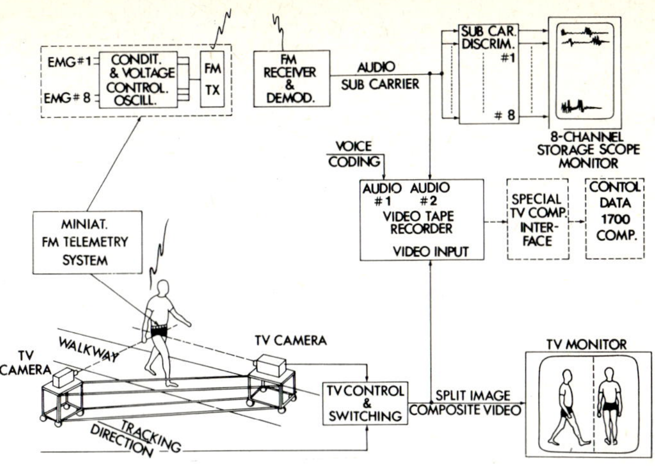

Fig. 1.4. Modern technology using a computer for studying human gait. From R. M. Letts, D. A. Winter, and A. O. Quanbury (1975), *Canadian Medical Association Journal*, **112**, 1091–1094.

computer print-outs from converted television fields (fig. 1.5). In fig. 1.5 only one-bit resolution is required to distinguish the leg markers on the subject; the contour of the leg has been drawn in to indicate their positions. The larger background markers serve as absolute reference points in the plane of progression. The present extensive use of computers in research has often changed problems from how to collect information to what to do with vast accumulations of it, the *reduction of data*. So many things can be measured so often that the measurements must be drastically telescoped so that researchers can figure out what they mean.

Currently there is also research interest in the gaits of horses, which represent big business in racing in many countries. At the University of Queensland, Australia, films of racing horses taken at 400 frames per second are studied frame by frame to detect subtle abnormalities not evident to the naked eye. In this way reasons for a horse's poor performance can often be pinpointed and rectified.

In general, the study of the locomotion of quadrupeds has progressed only recently, and at a much more leisurely pace, than that of man's, because such work is usually of academic rather than of practical value. Although zoologists have undertaken research in Great Britain, Canada, United States and Russia on a range of species, most wild animals have not been studied at all, and none has been studied to anything like the extent that man has been. With the present research

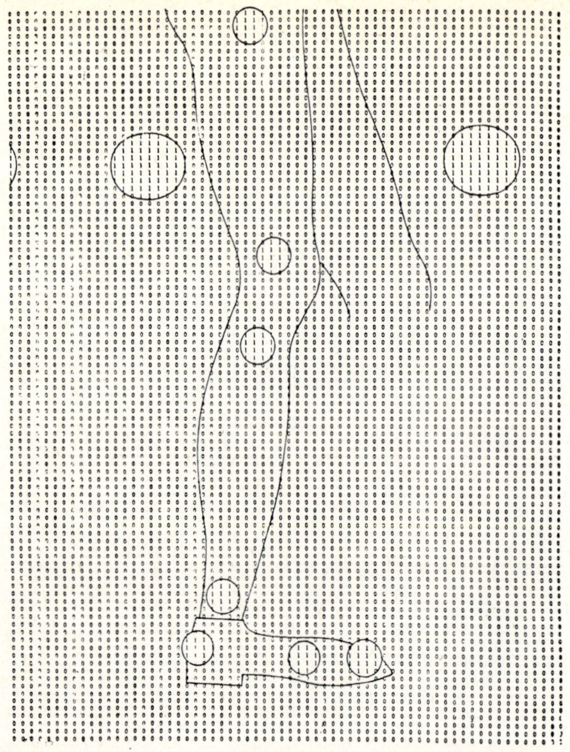

Fig. 1.5. Computer print-out of a converted TV field obtained as illustrated in fig. 1.4. The small circles indicate the position of markers fastened to the leg. From D. A. Winter, R. K. Greenlaw, and D. A. Hobson (1972), *Computers and Biomedical Research*, **5**, 498–504.

priorities and restriction in funds for scientific study throughout most of the world, it will be a long time before we know a great deal about the locomotion of wild animals.

1.6. *Current Concepts*

Despite the research that has been done on the locomotion of animals, little information has filtered through to the average zoologist or to laymen. Many authors still write about camels trotting, or lions pacing, or kangaroos bounding, verbs which are either misleading or wrongly used to describe the gaits of these animals. Since the advent of photography there is no reason why moving animals should not now have their legs drawn correctly, but errors are still prevalent. The brown bear *Ursus arctos* in fig. 1.6 was copied from a picture in a recent book on mammals, drawn as no bear has ever walked. The left hind foot is still

Fig. 1.6. Incorrectly drawn brown bear copied from a recent book on mammals. The left hind leg is being left behind, and the right hind leg has nowhere to go.

on the ground as the left front foot is being set down, so the bear's stride is much smaller than it should have been. Although the right hind leg should be the first leg to be moved on the right side of the bear, it obviously has no place to go to, given the artist's concept of the bear's stance.

2. Adaptations of Locomotion

2.1. *Use of Locomotion*

All animals use some form of locomotion. Among lower forms it is one of the criteria used to help distinguish animals from plants; one-celled forms that swim are usually classed as animals, those that do not as plants. Because this book is concerned with running, walking, and jumping, it is devoted to animals with legs. If an animal has legs, its trunk can move forward at a fairly constant speed, unhampered by contact with the ground. Meanwhile the legs, which do touch the ground, undergo large fluctuations in velocity. Since the energy wasted by these need not be supplied to the trunk but only to the legs, which make up a small proportion of the mass of an animal, there is a substantial saving of energy.

Just because an animal has legs, this does not mean that it travels to any extent. The locomotion of a group such as the sloths is minimal. These animals hang upside down in trees, moving slowly along the branches to find fresh leaves to eat. They venture to the ground perhaps once a fortnight to defecate and leave scent marks for communication with other sloths. They are unable to stand normally or to walk on the ground. Most animals with legs are less lethargic than sloths, but almost all still spend their lives in a restricted area. We like to think of wild animals as free to go where they like, but in fact they seldom venture outside a limited home range. A giraffe, for example, could travel the length of Africa many times in its lifetime, but it never does so. If it has ample food and water it will remain within one 100 km^2 area as long as it lives. Only migratory species such as wildebeest and caribou habitually cover great distances in their search for food and water.

2.2. *Maximum Speeds*

Cursorial or running animals are adapted not to a specialized habitat such as water, trees or earth, but to the more or less open country. They are all at least of medium size or larger, because they cover large distances in escaping from predators or catching prey. They seldom have burrows or nests in which to rear their young, although hunting canids and hyenas do. All cursorial animals are fast runners, as table 2.1 shows, but their speed is not correlated with their size except in small species (fig. 2.1). The fastest runners such as the cheetah and the pronghorn are relatively small. Very large animals cannot run especially fast because they have heavy bodies to transport using long muscles which do not contract as quickly as short muscles. A jackal can run faster than a giraffe even on its short legs because it can move them so quickly.

TABLE 2.1. *Maximum speeds of animals.*†

Animals	Speeds in km h^{-1}	Animals	Speeds in km h^{-1}
Cheetah	112*	Mule deer	56
Pronghorn antelope	98	Jackal	56
Wildebeest	80	Reindeer	51
Lion	80*	Giraffe	51
Thomson's gazelle	80	White-tailed deer	48
Ostrich	80	Wart hog	48
Quarter horse	76	Grizzly bear	48
Wapiti	72	Cat (domestic)	48
Cape hunting dog	72	Man	42*
Coyote	69	Elephant	40*
Gray fox	67	Squirrel	19
Hyena	64	Pig (domestic)	18
Zebra	64	Fowl (domestic)	14
Mongolian wild ass	64	Spider	1·9*
Greyhound	63	(*Tegenaria atrica*)	
Whippet	57*	Giant tortoise	0·3*
Rabbit (domestic)	56	Three-toed sloth	0·2*

†Most of these speeds were measured over a 0·4 km distance. Those marked with an asterisk were measured over a shorter distance.

Three bipedal species are included in fig. 2.1. Man is much slower than quadrupeds of the same mass, even though the human speed cited is for a champion and far higher than the speed an average man or woman could attain. The domestic fowl is slow too for its size, but the ostrich is very fast. It is impossible to generalize that quadrupeds go faster than bipeds of the same mass, even though they have twice as many legs.

2.3. *Is Fast Best?*

Often tables such as 2.1 leave one with the impression that most animals have evolved or are evolving to be as fast as their size and shape allow them to be; whether predator or prey, they will then be better able to survive. This is not necessarily true. D. Mech, for example, found that in winter on Isle Royale in Lake Superior those moose which tried to run from attacking wolves were more likely to be killed than those which stood their ground. There was thus no natural selection for speed in the moose.

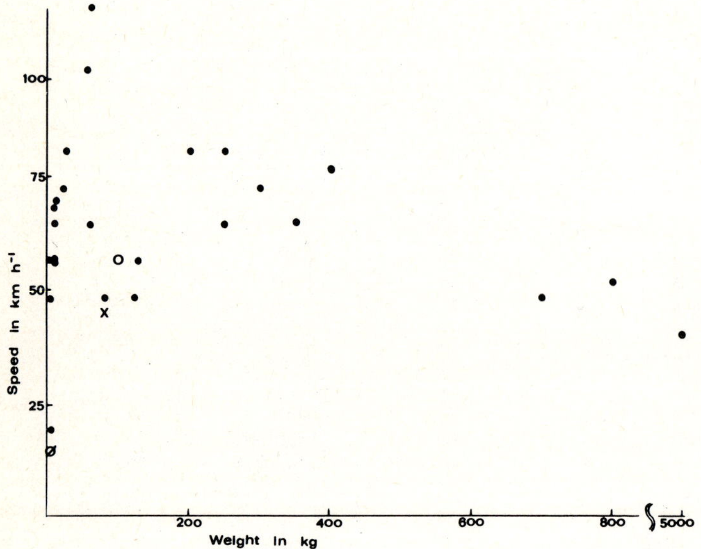

Fig. 2.1. Speeds of mammals and birds plotted against their masses. The fastest animals are medium-sized. Large animals have long legs, but also heavy bodies; small animals are light, but have short legs. The quadrupeds are marked with a ● ; man × ; ostrich o; fowl ø.

For wildebeest in East Africa the selection pressure of predators such as lion, cheetah, and hyenas is apparently much greater on calves than on adults. Slow calves are often killed, but so are calves that stray momentarily from their mothers, and calves that are less precocious than average. Since there is no evidence that slow calves grow into slow adults, there is no reason to believe that the fastest wildebeest survive to breed more often than their slower brothers. Observations in the field indicate that predators tend to be attracted to wildebeest that look or behave differently from the other members of the herd. One zoologist found to his distress that individuals he had marked with paint on their horns so that he could identify them were pursued and killed by lion more often than unmarked animals.

2.4. *Is Locomotion Limiting?*

Visitors to the Canadian arctic have commented on the number of limping caribou they have seen there, presumably animals that have fallen on the uneven terrain and sprained or broken a leg. If locomotion were limiting, one would imagine that such animals would quickly be taken by wolves, and perhaps they are; wolves have trouble catching healthy caribou. Animals which are large, and therefore have few predators, are certainly not limited by their locomotion. Two disabled giraffe survived well in the wild, one which limped because a foot had become caught in a wire noose, and another which was without the distal end of one of its front legs.

Small prey animals are generally limited by the way they can move; probably few injured lizards or squirrels or rabbits survive long in the wild. Small carnivores are more fortunate. M. Taylor found that 15 per cent of 308 individual viverrids examined by him had some pathological condition because of earlier fractures or diseases which would certainly have affected their movements. Yet these individuals had survived their ailments and lived to hunt again.

Birds too can survive despite leg injuries, because they can still fly from danger. One silver gull which had a leg bitten off by a fish was able to survive well at an Australian resort, hopping aggressively among other gulls to snatch morsels of food. A one-legged crow in the Australian desert was also seen over a long period of time, apparently leading a normal life.

2.5. *Endurance*

Cursorial animals are adapted for speed, but sometimes also for going long distances. Such animals are always big ungulates. With their long legs they can travel with a minimum expenditure of energy per unit body mass and they need relatively less food than smaller animals, even though they are physically able to collect more. Large animals have relatively few predators; they also live long enough to experience many things and to pass on what they have learned to their offspring.

The animal with the greatest endurance is the dromedary, which may spend its life wandering in the Sahara desert, hunting for edible vegetation. Even in hot weather it can last for weeks without water, drinking up to 200 litres when it finally reaches a well. Camels seldom use the tiring gallop, having perfected instead the pace, a relatively fast gait which they can keep up for many kilometres.

Caribou have less endurance than dromedaries, but they also travel thousands of kilometres each year. In North America they go north each spring to the rich tundra meadows, and south each fall to pass the long winter in the taiga. Like camels, caribou seldom gallop unless they are chased by men or wolves. Their fast gait is usually the trot, which enables them to cover long distances without tiring.

A third migratory animal is the wildebeest of the Serengeti in Africa. The herds travel long distances each season so that they can take advantage of fresh young grass which appears on the savannah following heavy rains. Like the other species, wildebeest migrate by walking if there are no predators to be evaded; if there is danger, only the animals nearest it may run, while those further away plod on unconcerned. The fast gait of the wildebeest, as of the small saiga and pronghorn antelopes which also travel long distances in search of food, is the gallop, a gait relatively less tiring in small than in large species.

2.6. *Other Adaptations*

The cheetah and the pronghorn antelope are examples of plains animals which have evolved as fast cursorial species. Other plains animals have evolved somewhat differently. The springbok and impala rely not only on speed in flight, but on erratic high jumps and sudden changes of direction. Lions put all their effort into short sprints, because they do not have the stamina to run far. Leopards and baboons are fairly fast, but they have also the agility to climb trees. The evolution of any form has been shaped by a number of factors, only one of which may have been for increased speed. Speed in an animal may be important, but factors such as camouflage, behaviour, and habitat may make it secondarily so.

3. Morphology of Vertebrates

3.1. *Basic Materials for Locomotion*

Animals with legs progress by contracting various muscles which move various leg bones as levers about various joints. Individual muscles can be either large, as our *gluteus maximus* or rump muscle is, or minute, as are our toe muscles. The voluntary or striated muscles are composed of many cylindrical syncytia (not cells) called fibres, each from 50 μm to 50 mm or more in length. Man has about two million of these, which run parallel to each other to form muscle tissue. Muscle fibres are bound with connective tissue sheaths into fascicles. It is this connective tissue, not the muscle fibres themselves, that is connected directly to cartilage or bone. If there is little such connective tissue the muscle assumes a fleshy attachment, but if the connective tissue is extensive it forms either a long *tendon* or a wide *aponeurosis* or *fascia*. The *origin* of a limb muscle is the relatively stationary proximal end nearest the trunk, and the other end, the *insertion*, is further down the limb.

The force of a muscle can only be exerted through a contraction of its fibres. This force is proportional to a muscle's cross-sectional area, which is the reason that weight-lifters need immense arm and leg muscles. The distance it can contract, nearly half its length, is proportional to its length. However, the microstructure of a muscle must also be taken into consideration in analysing how the muscle works. Muscles composed largely of white fibres contract faster and tire sooner than those composed largely of red fibres.

The action of shortening a muscle and pulling together the bones or other structures to which the two ends are attached is triggered by a motor nerve fibre. This nerve joins the muscle at a *motor end plate* or nerve–muscle junction on the outer surface of a striated fibre. Each such nerve usually divides terminally to supply a number of adjacent muscle fibres.

The muscles of locomotion can be divided into *extrinsic* muscles, which arise from the body and run to the pectoral girdle, the pelvic girdle, or a limb (e.g. the *psoas* muscle); and *intrinsic* muscles which originate and insert entirely on one limb (e.g. the *gastrocnemius* muscle). Depending on how a muscle moves a leg or part of a leg, it is an *extensor* if it acts to increase the angle between two limb segments; a *flexor* if it decreases it; an *adductor* if it draws a segment towards the body; an *abductor* if it pulls a segment away from the body; a *protractor* if it moves a limb forward; a *retractor* if it moves a limb backward; a *levator* if it raises a structure; a *depressor* if it lowers one; and a *rotator* if it twists a limb segment. Usually movements are accomplished by the action of two or more muscles working together as *synergists*. Muscles cannot actively elongate. When a movement is completed, a muscle stops contracting and is usually pulled out by the action of an *antagonistic* muscle.

All movements take place where the bones meet at the joints. These have a variety of structures which allow either small or large movements of the bones in one or more planes. In man the shoulder joint has the largest range of movements, whereas the vertebral joints allow the spinal column to bend only slightly.

The skeleton, which forms the framework of the body on which the muscles act, must be strong enough to support the standing weight of an animal, plus the forces exerted by that weight under large stresses. Thus a 70 kg man when running exerts a stress of more than 35×10^6 N m^{-2} on the shaft of his femur. Such stresses on a bone may be reduced if muscles or tendons adjacent and parallel to it are taut enough to take some of the stress from the bone.

The vertebrate skeleton is divided into *axial* parts (skull, ribs, vertebrae) and *appendicular* parts (pectoral and pelvic girdles, limb bones). The forelimb is joined to the trunk by the pectoral girdle, which in tetrapods is anchored to the axial skeleton by muscles and connective tissue, not bone, so that there is some spring in the suspension of the forequarters of the body. This is important in galloping and bounding, because an animal absorbs the shock of impact on its front rather than its hind legs. The stronger pelvic girdle articulates with or is fused to the vertebral column, so that any impetus from the hind legs is passed directly on to the trunk.

3.2. *Evolution for Improved Locomotion*

In the evolution of the vertebrates one important change has been the displacement of the legs from supporting the body from either side (in amphibians and living reptiles) to supporting the body from underneath (in birds and mammals), a much less tiring and more efficient arrangement (fig. 3.1). The mammalian limbs have evolved elbow joints which point backwards following progressive torsion of the limb, and knee joints which point forwards principally because of changes in the

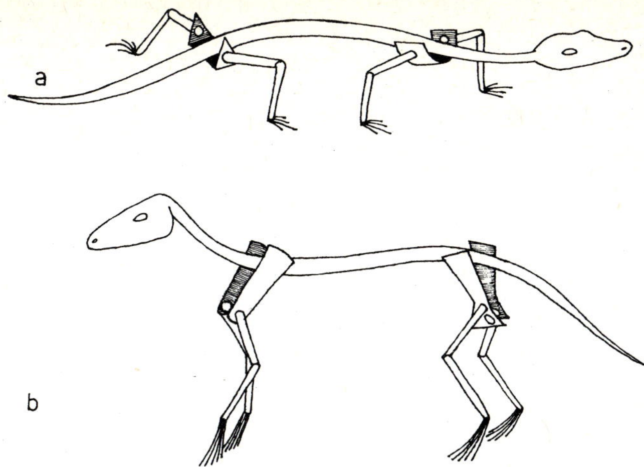

Fig. 3.1. Representative stances of (*a*) a reptile, and of (*b*) a mammal. In modern reptiles the legs are lateral to the trunk, and in mammals they are under the trunk.

anatomy of the hip joint. The limbs thus move in a rough parasagittal or anterior–posterior plane, more or less rigidly depending on the species. Small mammals tend to have unspecialized limbs because they rely for protection on burrows from which they seldom venture far. Larger terrestrial mammals however are often specialized for *cursorial* (running), *saltatorial* (jumping), or *graviportal* (heavy-weight bearing) locomotion. Some small mammals also are saltatorial.

Evolution in cursorial animals has emphasized leg rather than body movements, in particular increasing leg lengths so that strides will be longer. One way to increase the length is to change the stance from *plantigrade*, in which the whole foot comes in contact with the ground, to *digitigrade*, in which the digits support the legs, to *unguligrade*, in which only the tips of the toes touch the ground (fig. 3.2). Thus a marker such as man's ankle or knee moves higher up the leg in more cursorially evolved mammals. What seems like the knee in a horse is not comparable to man's knee at all, but to his ankle. Two main groups of mammals use the unguligrade stance, the Order Perissodactyla (horses, rhinoceros, tapirs), whose weight is supported through the third digit with the others variably reduced, and the Order Artiodactyla (four- or two-hoofed ungulates) whose weight is carried through the third and fourth digits. Some species such as bears have plantigrade hind legs but digitigrade front legs, while others such as man change their support stance when they run from plantigrade to digitigrade. The type of stance can be determined by examining footprints (fig. 3.3).

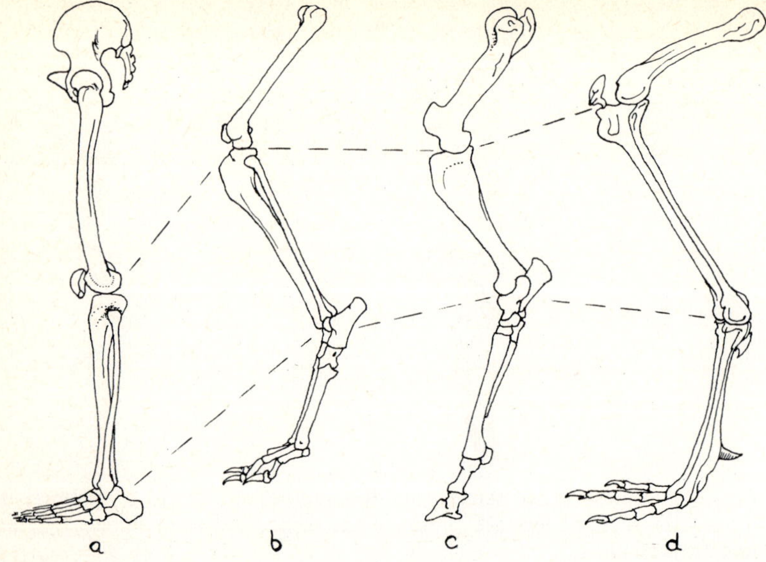

Fig. 3.2. Hind legs showing variable elongation. (*a*) Plantigrade (man); (*b*) digitigrade (dog); (*c*) unguligrade (horse); (*d*) bird. Not to scale.

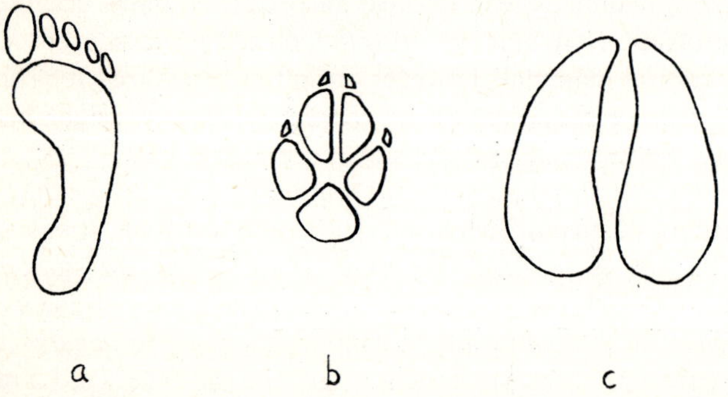

Fig. 3.3. Footprints of (*a*) a plantigrade (man); (*b*) a digitigrade (fox); and (*c*) an unguligrade (deer). Not to scale. The stance can be inferred from the print.

Cursorial animals often achieve increased leg-length not only because of their stance, but through an absolute increase in the length of their leg bones. The humerus and femur remain relatively unchanged from the ancestral condition, but the other leg bones increase in length progressively from proximal to distal, including the foot bones. With the loss (in ungulates) or reduction (in carnivores) of the clavicle, the shoulder blade becomes free to swing along with the attached foreleg. Because the leg now pivots above, rather than at, the shoulder, this change further increases the length of an animal's stride.

The limbs of cursorial animals act very much like driven pendulums, with as many joints as possible moving in the same direction at the same time. Simplification in the movements of the joints means that the amount of musculature needed in the leg is decreased, and with a less massive lower leg and smaller moment of inertia, theoretically less energy should be associated with changes in the leg's angular momentum as it swings back and forth. (This decrease was not found experimentally to make much difference, though, when the energetic cost of running of cheetahs, with heavy legs, was compared with that of goats and gazelles, with slender legs.) The mass of the lower leg is also reduced by having tendons function here, with the attached muscle tissue concentrated close to the trunk where the fluctuations of velocity are least. The muscles which flex and extend the knee joint could function in the calf region, but they are found instead in the thigh.

In cursorial mammals there has also been a reduction of bone elements, with the basic five-digit limb of vertebrates reduced to one or two sets of elements in the most evolved ungulates (fig. 3.2). A foot with one or two toes is lighter than one with more, and a single cannon bone in ungulates is preferable to five metacarpals or metatarsals.

Some cursorial mammals such as cheetahs or greyhounds are fast because they have evolved both long legs and supple vertebral columns. As they launch outstretched into the air, their backs are extended too, thus increasing the length of the stride. As well, extra rotation of the hip and shoulder girdles in the direction of the swinging legs added to the flexion and extension of the spine increases the swing of the legs themselves, so that they reach out farther front and back than they could if the spine were rigid, and strike and leave the ground at more acute angles. M. Hildebrand has calculated that the cheetah's spine is supple enough to increase its stride length to such an extent that theoretically it could run at nearly 10 km h^{-1} without any legs at all!

3.3. *Bones and Locomotion*

Bones are so vitally involved in all types of locomotion of vertebrates that often much can be inferred about the way a species moves from its skeleton. As an extreme example, the fact that man walks on two legs is associated with radical modifications of his skeleton from the basic quadrupedal form of mammals. These changes include:

(1) A skull centred on the vertebral column, with the foramen magnum and the occipital condyles which connect the skull to the backbone under, rather than at the back of, the skull. The face and

jaws are retracted and thus less able to gather food, but the hands have taken over this function.

(2) A broader thorax which thus lies more nearly over the centre of gravity, which runs just anterior to the vertebral column.

(3) A curved backbone which decreases the shock to the brain of successive foot-impacts during walking and running.

(4) A broad pelvis which supports the abdominal viscera to some extent and to which the heavy leg and trunk muscles are attached.

(5) Knee joints which readily lock when extended, so that a person may stand for a long time with little strain on the postural muscles.

(6) Legs which attach widely to the pelvis, yet approach each other fairly closely at the knees in a standing position, so that the centre of gravity runs through them from the trunk to the ground.

(7) Feet which serve both as pedestals and for simple walking. The bulky calcaneum is elongated, the tarsals are compacted, and the metatarsals which are often splayed for manipulation (as in cats) have been brought into sagittal line. Special muscles and stays bind these bones together into springy arches.

If we had only man's skeleton to examine, as we have only the skeleton of our extinct relative *Australopithecus*, we could still postulate a great deal about such an animal's movements. Indeed palaeontologists have hypothesized how *Australopithecus* moved from its pelvis alone (fig. 3.4). The ilium of *Australopithecus* resembles that of man much

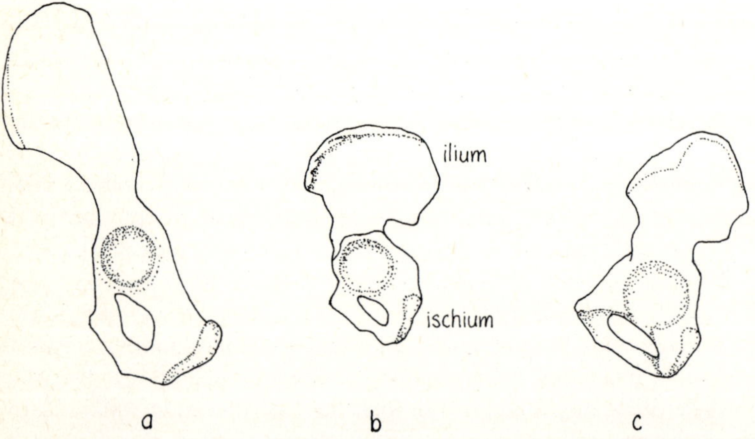

Fig. 3.4. Comparison of the pelvis of three higher primates. (*a*) Gorilla; (*b*) *Australopithecus*; (*c*) modern man. Front is to the left. Only that of modern man allows him a long stride. Based on work of J. Napier, *Scientific American*, April 1967, pp. 56–66.

more closely than that of the gorilla, which is a quadruped, so it is assumed that *Australopithecus* was bipedal. However, its ischium is longer than man's, and prevented it from taking long strides in walking or running. For this the leg has to swing back to a position behind the vertical axis of the spinal column, a degree of extension that can only be achieved if the ischium is very short. The pelvis of *Australopithecus* also shows that the gluteus medius and the gluteus minimus muscles, which in man are of primary importance in stabilizing the pelvis during each stride, were less well developed. Apparently this hominoid covered the ground with quick, rather short, steps, with its knees and hips slightly bent. There was no possibility of a prolonged stance phase in walking as is found in man, nor of *Australopithecus* travelling large distances using its tiring walk.

Despite the valuable information that can be gleaned from the anatomy of a species' bones, care must be taken not to assume too much. Thus one might judge from their enlarged hind legs that the pocket mouse *Perognathus* and the jerboa marsupial *Antechinomys* hop bipedally, but such an assumption would be wrong, as we shall see in Chapter 8.

The study of bones and their relation to locomotion need not be a boring 'dry-bones' kind of exercise, especially now that cineradiography is being used to study how the bones of an animal actually move during locomotion. F. A. Jenkins of Harvard University has taken X-rays of a walking echidna (*Tachyglossus aculeatus*) and concluded from them that this monotreme's limbs are not reptilian in posture and movement as had been assumed. Instead the forefoot was planted on the ground under the shoulder rather than lateral to it, and the movement of the femur was similar to that of generalized mammals. Jenkins used the same cineradiographic technique to compare the limb movements of seven other small mammals (fig. 3.5). Only the cat and, to a lesser degree, the hyrax (*Heterohyrax brucei*) had the postural and limb movement pattern traditionally regarded as typically mammalian, with the legs directly under the body and moving in one plane back and forth; in the opossum (*Didelphis marsupialis*), tree shrew (*Tupaia glis*), hamster (*Mesocricetus auratus*), rat (*Rattus norvegicus*), and ferret (*Mustela putorius*), the humeri and femora usually functioned in positions more horizontal than vertical and at angles oblique to the parasagittal plane. If there exist living mammals that retain the postural and locomotory patterns of their Mesozoic ancestors, they are non-cursorial species such as these five.

Bones can also be tested for strength, to determine how much stronger they are than they need to be for movements in any particular species. L. Calow and R. M. Alexander studied the femur of the common frog, *Rana temporaria*. When the bone from a freshly killed frog was held at one end and pushed sideways at the other, it took a bending moment of about $2 \cdot 6 \times 10^{-3}$ newton metre to bend it through

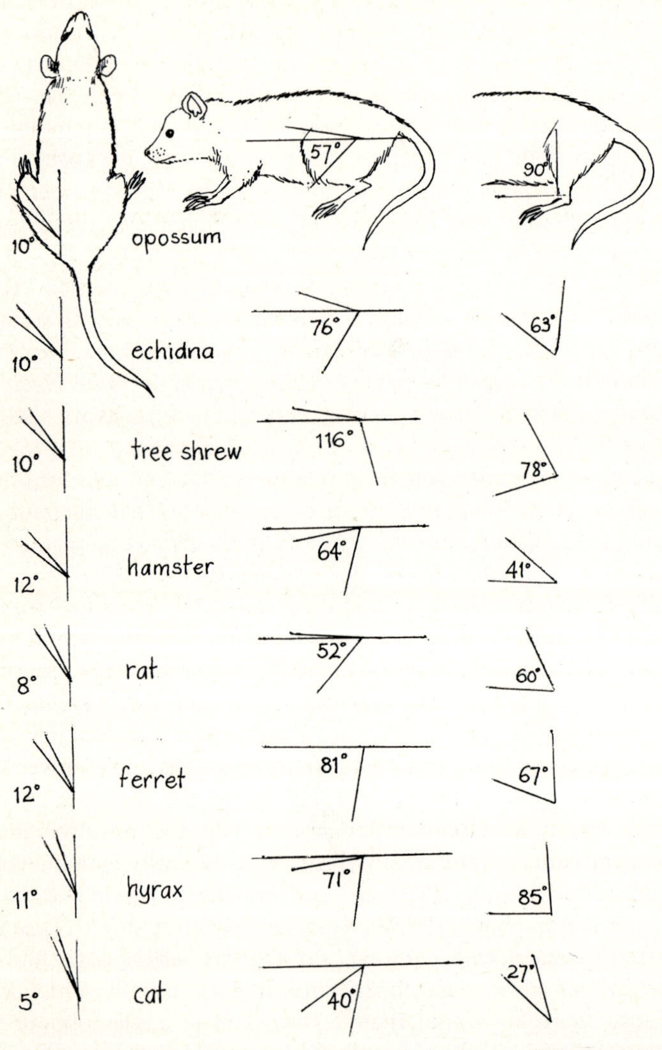

Fig. 3.5. Comparison of hind limb movements expressed as excursion arc. (*a*) Movement of femur relative to parasagittal plane; (*b*) Movement of femur relative to horizontal; (*c*) Movement of tibia and fibula relative to horizontal. Based on the work of F. A. Jenkins, Jr. (1971), *Journal of Zoology* (London), **165**, 303–315.

one degree. A bending moment of 63×10^{-3} newton metre had to be applied before the bone broke, usually through its shaft. When a frog jumped strongly from a force platform, a ground reaction of about 0·35 newton was recorded. Taking the moment arm of this force as half the length of the 25-mm femur, the greatest bending moment which joint forces are likely to exert on their shafts during jumping is about 5×10^{-3} newton metre, enough to bend them through only 2–3°. Thus the femur is far stronger than it would apparently need to be.

3.4. *Muscles and Locomotion*

Muscles, which work with bones to produce all the movements of which vertebrates are capable, can be studied in relation to locomotion in several ways. In comparative studies dealing with a wide range of species, researchers have measured the volume of various muscles, on the theory that muscles which are used a great deal will be larger than those not used much and therefore less developed. Such a study on birds showed, as might be expected, that strong fliers had large pectoral muscles, while cursorial birds that seldom flew had large leg muscles. However, such comparisons do not take account of important proper-ties of muscles such as their make-up of red and white fibres, their angle of action across joints, and variations in their shape. Moreover, muscles do not work alone but in different combinations, and their actions may change at different times and under differing conditions. Even if a muscle is reduced in one species, in another its action may be more than compensated for by synergistic effects or by increased leverage caused by changed shape and insertion. Comparative studies of the musculature of closely related species are likely to yield informa-tion more important in taxonomy than in locomotion.

Some workers have studied muscle development and locomotion in living animals, but such research should be discouraged because it causes the subjects pain without yielding data of much practical value. One study by Dutch and English workers showed that muscle weights could reflect differing muscle use in individuals of the same species. They cut off the tails and three-quarters of the forelimbs of 31 three-day-old laboratory rats. These rats as they grew were as active as their normal littermates, either walking or hopping on their hind legs, or sliding along using their forearm stumps and hind legs in sham-walking. At the age of 4 months or 7 months the experimental and control rats were killed, and the muscles of their hind quarters weighed. Those of the bipedal rats were relatively heavier than those of the normal rats because of their increased use during the rats' development, with individual muscles more or less hypertrophied depending on their importance in bipedalism.

3.5. *Action of Muscles on Bones*

A classic comparative study in anatomy was undertaken in the 1950s by J. M. Smith of University College, London, and R. J. G. Savage of the University of Bristol, who compared the forelimb and scapula of the horse (*Equus caballus*), a cursorial species, and the armadillo (*Dasypus novemcinctus*), a fossorial (digging) one. The horse is adapted for fast running, so the front legs must be able to move quickly back and forth; the armadillo need move its front legs only slowly, but it must be powerful enough to overcome resistance of the earth during digging. One of the main muscles flexing the forearm in both species is the teres major which originates on the scapula and inserts on the humerus (fig. 3.6). We can compare the action of this muscle with

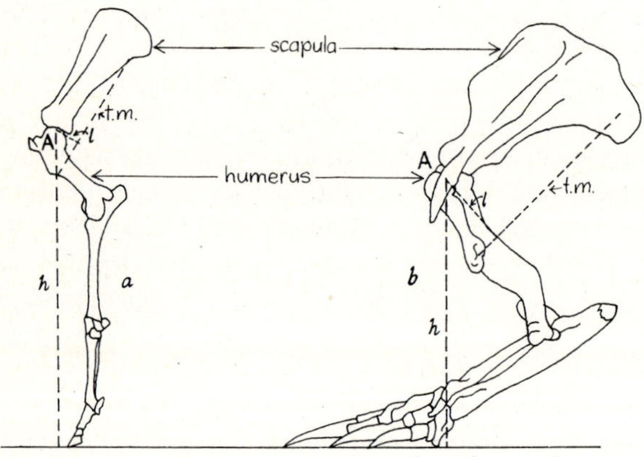

Fig. 3.6. Left front legs of (*a*) the horse and (*b*) the armadillo. The line of action of the teres major muscle (t. m.) running between the scapula and the femur is indicated. A is the joint; *l* the distance between A and the line of action; and *h* the distance of A from the ground. Based on the work of J. M. Smith and R. J. G. Savage (1956), *Linnean Society* (London) *Zoology*, **42**, 603–622.

swinging on a gate. If we sit near the hinge, we have less speed as we swing than if we sit near the end, since the speed is proportional to the distance from the hinge. Similarly if v is the rate of contraction of the teres major muscle, V the speed imparted to the foot relative to the shoulder girdle, l the perpendicular distance from the glenoid joint (hinge) to the line of action of the muscle (= moment arm of the muscle about the fulcrum A), and h the perpendicular distance from the glenoid joint to the ground (fig. 3.6),

$$\frac{v}{V} = \frac{l}{h}, \text{ so } V = v \times \frac{h}{l}.$$

To consider the forces involved we can refer again to the gate. It takes a larger force to close the gate if we push near the hinge than if we push near the edge, and our force is inversely proportional to the distance from the hinge at which we apply it. Similarly, if F_1 is the tension in the teres major muscle (and hence the force exerted at each end of it) as it contracts, and F_2 the backward thrust produced on the ground by the leg,

$$F_1 \times l = F_2 \times h, \text{ so } F_2 = F_1 \times l/h.$$

Because the value of l/h in the armadillo is relatively large ($\frac{1}{4}$), its movements are powerful but slow; because the value in the horse is small ($\frac{1}{13}$), its movements are fast but weak. In the horse the scapula is long and held nearly vertical, so the teres major is long too. A contraction of $\frac{1}{5}$ its length would rotate the humerus through $50°$, and this could be done quickly because of the short moment arm ($= l$ about A). In the armadillo the scapula is directed at about $45°$ to the horizontal, so that a contraction of the teres major by $\frac{1}{5}$ its length would rotate the humerus through an arc of only about $30°$. This would be done fairly slowly, as the moment arm here is relatively much longer. In general all digging animals have short scapulae with large scapular spines, while running ones have long scapulae, with the spines less developed in carnivores and weak in ungulates.

3.6. *Nervous Control of Locomotion*

Physiologists study locomotion by way of the function of muscles and nerves, a more difficult and recent approach than that of gross anatomy, and one so far dealing more often with amphibians than with the more complex mammals. Great progress has been made at the cellular level in investigating the contractile mechanism of the muscle fibre, the biochemistry of the motor end plate, and the transmission of the nervous impulse, but understanding of how these function together to make an animal move is much less certain. It is known that proprioceptive reflexes, or local reflexes of the spinal cord, play an important role in vertebrate locomotion. In cats whose spinal cords have been cut in the lower chest region, the lumbosacral cord can still coordinate alternate walking movements of the hind limbs. In similar experimental rabbits this part of the cord can produce not only alternate walking of the hind limbs, but also synchronous jumping movements.

Spinal reflexes in vertebrates are subordinate to the higher centres of the brain. In amphibians the total correlation of locomotion is centred in the mid-brain roof (*optic tectum*); the mid-brain floor contains the motor centres which have general control over all those of the spinal cord, with input from centres in the hind-brain and the palaeocerebellum. In diapsid reptiles and birds the total correlation for movement is shifted forward to the fore-brain, with the mid-brain roof remaining as a subordinate correlation centre. In mammals this correlation centre is dominated in turn by a third development of a total correlation centre, the cerebral cortex, which has an enormous capacity for growth in absolute size and in the number of neuronal connections, with consequent great possibilities for storing information, for learning, and for developing new patterns of behaviour. One German Alsatian dog was still able to function after it had accidentally lost both legs on one side. It could spring up from the ground and gallop along on its two legs, but it collapsed when it stopped and it had no slow speed. Similarly a young goat born without front legs was able to move about by standing on its hind feet and hopping. If the brainstem of a cat is transected so that only the mesencephalon remains, the animal can make walking movements if a circumscribed region below the inferior colliculus is stimulated repetitively. If the stimulation strength is increased, the speed of walking increases, with the diagonal limbs gradually acquiring a greater degree of synchrony, as in a trot. With a further increase in stimulation strength, the pattern changes to that of a gallop.

The great variety of movements possible in mammals contrasts with the few present in the small-brained amphibians. Movement of amphibians' legs is reflex, and the muscles are activated in an invariable, ordered sequence to produce locomotion. Reflexes of this sort are apparently phylogenetically old and genetically determined. The evolution of neuronal pathways in higher vertebrates, which are as important to locomotion as the evolution of limb morphology, has been little studied, although I uncovered one possible approach when analysing the swimming styles of various mammals. Both rabbits and chinchillas hopped on land, but rabbits usually swam by moving their legs alternately. Chinchillas by contrast only made hopping movements in water with their front and hind legs moving synchronously, a gait so ineffectual that they would soon have drowned had I not rescued them. Perhaps the hopping gait is phylogenetically so old in chinchillas that they can move their legs in no other way, while the hopping of rabbits is relatively recent and can on occasion be replaced by walking, as when they are very young, or must navigate in mud or deep snow where hopping is ineffective.

4. Human Walking and Running

4.1. *Bipedal Locomotion*

In the course of history locomotion on two instead of four legs has evolved at least six times among vertebrates. All of these animals had big feet, so that they could balance while standing without undue muscular effort. (A person walking on stilts or one manoeuvring on a tight-rope has problems of balance which accompany a small base of support.) Bipedal vertebrates can move in five main ways:

(*a*) The alternate use of the hind legs which is employed by various primates, including man, and in which the body is balanced vertically, or nearly vertically, over the hind limbs. This bipedalism apparently evolved because of the need in some primates to use their front legs for other things beside locomotion, such as gathering food and carrying loads.

(*b*) The alternate use of the hind legs common to most birds in which the body is not balanced vertically, but more nearly horizontally, with the centre of gravity directly over the hind legs. A bipedal gait is essential in birds, where the forelimbs have usually become entirely adapted for flight.

(*c*) The alternate use of the hind legs in which the body is balanced more or less horizontally over the legs and the tail acts as a cantilever. This reptilian mode of locomotion probably evolved at least three times, in two groups of dinosaurs and in the lizards, apparently because of the need for speed.

(*d*) The synchronous use of the hind limbs in the ricochet. This method of progression has evolved in both marsupial and placental mammals. As in most primates, the saltatorial mammals also have their bodies balanced semi-vertically over their hind legs.

(*e*) The ricochet is used also by small birds and some larger ones, the body being carried in a similar way to that of those that walk.

4.2. *Primate Evolution*

Primates and their type of locomotion have evolved over millions of years, beginning with arboreal forms which had acquired, as a result of

their habitat, binocular vision, a high degree of neuro-muscular
co-ordination, prehensile extremities, elongated forelimbs, increased
mobility in the elbow and shoulder joints, and a vertical posture. This
posture, evolved perhaps 50 million years ago, is evident throughout the
primates in all forms which sit on or hang under branches, or which
cling to tree trunks. Fossil primate species had skeletal adaptations
consistent with vertical clinging as early as the Eocene, and some
primates such as tarsiers still retain this habit, moving from one tree to
another by large leaps (fig. 4.1). By the Oligocene, hominoids and
monkeys had developed quadrupedal types of locomotion still predominant today with which they ran along branches of trees. Among
the hominoids brachiation, in which a primate moved about under

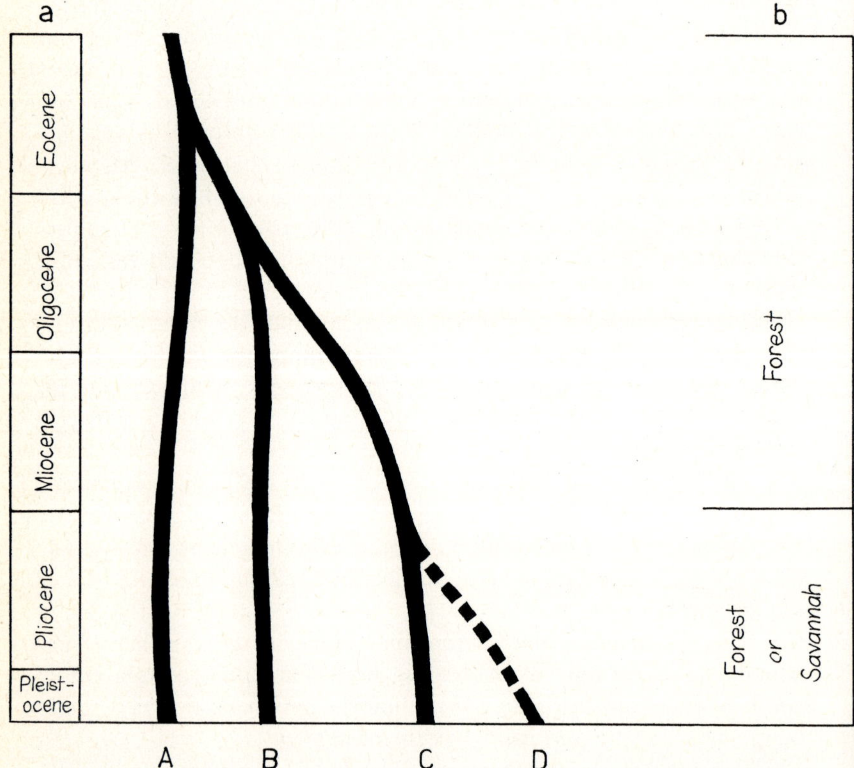

Fig. 4.1. Evolution of primates. (*a*) Locomotory trends in primates.
A. Vertical clinging and leaping (i.e. tarsiers); B. Quadrupedalism (i.e.
monkeys); C. Brachiation (i.e. gibbons); D. Bipedalism (man). (*b*) Habitat
of primates. Based on the work of J. Napier, *American Journal of Physical
Anthropology*, **27**, 333–342.

branches by swinging with its arms, first appeared in the Miocene. This type of locomotion was correlated with long arms, flexible shoulders, and sometimes an increase in size such as is found in orang-utans, chimpanzees and gorillas. The most proficient present-day brachiators however are the gibbons which have remained small.

Man himself probably evolved from a quadrupedal Miocene form such as *Proconsul* from East Africa, via a brachiating series of animals which eventually moved from an arboreal habitat to woodland savannah, to a bipedal creature which first scavenged and then hunted for its food. *Australopithecus*, as we have seen, was not perfectly adapted to bipedalism because it could not walk or run with full strides, but a million years ago a species of man had evolved which, judging by the anatomy of its foot, was as perfectly bipedal as is *Homo sapiens* today.

4.3. *Methods of Studying Human Gait*

When one analyses in detail the vast complexities of walking and running in man, one is amazed that almost everyone can do these things, and at an early age. To quantify even one stride of a moving person, one must measure and record for each part of the body continuous changes in speed, direction, rotation, and force. These values in a second stride often differ from those in the first, even if the subject is consciously trying to maintain a constant gait. Motion is restricted to joints, but more than one muscle crosses over each joint, so that the movement of bone levers caused by the contraction or relaxation of these various muscles can occur in a number of different directions at many different speeds. The methods described here were developed for studying human movements. Few of them have yet been used extensively on quadrupeds, although such research will be of interest when it is undertaken in the future.

(*a*) The most time-honoured method of studying locomotion is from *cinematograph film* often taken in slow motion. By looking at successive frames of a sequence one can determine the number of milliseconds that each leg spends on the ground and in swinging forward, and the relationship of one leg to the other(s). These data have been plotted for human beings (fig. 4.2), and also for other species, because the technique involved is simple and can be used on films of animals taken in the wild. In fig. 4.2, the proportion of a walking stride devoted by each leg in turn to the stance phase and to the swing phase can be readily seen.

(*b*) *Kinematics* is the analysis of the movements of bodies and parts of bodies in time and space, without considering the forces which cause these movements. Thus kinematic data can be gathered from specially

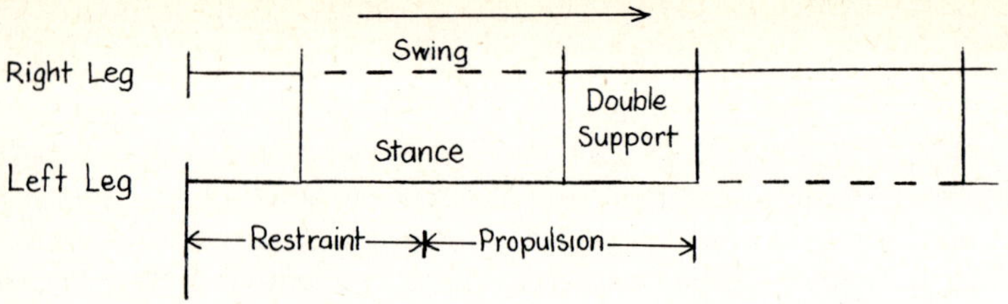

Fig. 4.2. Diagram of a human walk. When a foot touches the ground it first acts as a brake, reducing the forward momentum, then the reaction of the ground increases the forward momentum before the foot pushes off.

filmed sequences using body markers such as those of Marey to show how a bone or joint moves during a walking or running stride. From fig. 4.3 the spacing of the body markers with time (X-axis) indicates that the movement of the foot fluctuates much more strongly than that of the knee or thigh. There is also much more up and down movement in the ankle joint than in the joints higher up (Y-axis).

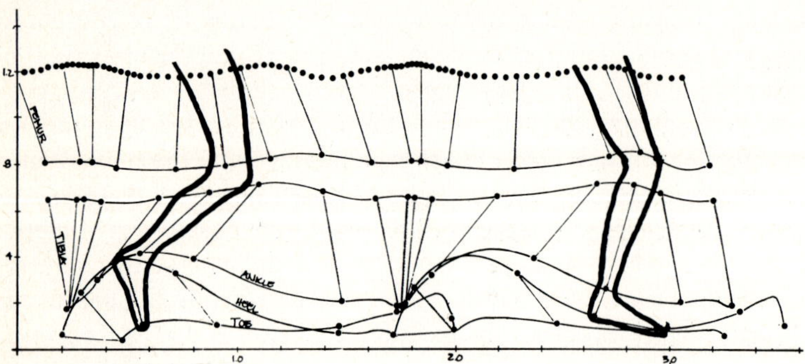

Fig. 4.3. Trajectory plot of spatial movement of body markers during the walk. The movement is less smooth the farther from the hip. From R. M. Letts, D. A. Winter, and A. O. Quanbury (1975), *Canadian Medical Association Journal*, **112**, 1091–1094.

Kinematic information can also be gathered by an *electrogoniometer*, first used in 1961, which is an instrument designed for the automatic recording of joint motion. The electrogoniometer consists of two arms, one of which can be attached to the calf and one to the thigh if the knee joint is to be studied, with three interposed potentiometers. One potentiometer can be oriented in each of the three planes of motion, so that movement of the goniometer arms will produce concurrent changes in the indications in the potentiometers. Different channels of a recording device can measure in degrees the flexion–extension of the joint, the abduction–adduction, and the rotation.

(c) *Kinetics*, which together with kinematics makes up the subject of *biodynamics*, studies the changes in motion caused by an unbalanced system of forces and determines what force is needed to produce a specific change. Kinetics is best studied using a *force-plate*, which measures during locomotion the ground reactions, and records directly the vertical accelerations. Essentially the apparatus consists of a heavy metal plate rigidly supported by metal columns to which are bonded a series of opposing pairs of strain gauges. The deformations along the x, y and z axes, as well as torque, are measured by balanced bridge circuits and recorded on an oscillograph. Some of this same information can be gleaned from the wear on the sole of a person's shoes. The wear pattern will reveal a limp that a woman is trying to hide, or unmask one that a man is trying to feign, perhaps to provide evidence that he was injured for an insurance claim.

(d) *Electromyography* is the subject which relates the movement of a body with the muscle action which causes this movement. Needle electrodes can be either inserted in specific muscles of the legs or attached to them, so that the exact time of contraction of each muscle can be correlated with the phase of the stride in an electromyograph (EMG). In fig. 4.4, which gives the tracings of EMG signals from five muscles during the normal walk plus the phase of the step (a), it is evident that as the heel touches the ground at the beginning of the stance phase (vertical line in fig. 4.4), the tibialis anterior and the vastus lateralis both contract, the former to flex the ankle and the latter to extend the knee. The gastrocnemius and soleus muscles which extend the ankle become active during the stance phase, reaching a peak of activity at heel-off just before the toe leaves the ground. The semimembranosus, one of the hamstring muscles, contracts during the swing rather than the stance phase, causing the hip to extend just before the heel touches the ground.

(e) Locomotion can also be approached from a consideration of gross morphology of animals, as in Chapter 3, and of energy relationships.

4.4. *Centre of gravity (centre of mass)*

Forward movement of an animal is complicated, because whereas the trunk moves more or less steadily along, the legs which cause its movement stop and start, alternately supporting the weight of the trunk and then swinging free, so that at any one time various parts of the body are moving at a variety of rates. The weight of a body, which is the

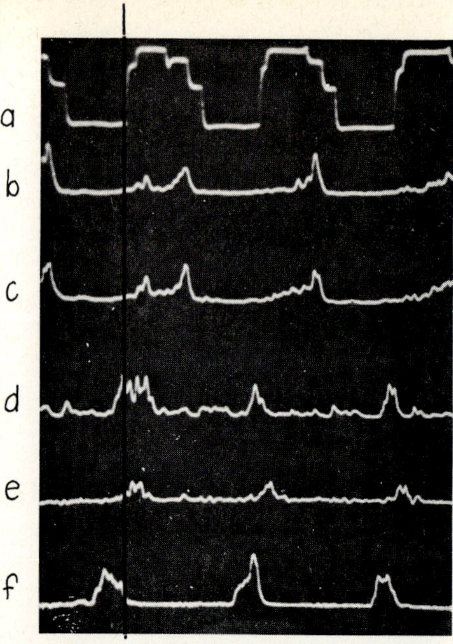

Fig. 4.4. Oscilloscope tracing of EMG signals from five leg muscles during a normal walk. (*a*) Stance and swing phase of leg; (*b*) Lateral head of gastrocnemius; (*c*) Soleus; (*d*) Tibialis anterior; (*e*) Vastus lateralis; (*f*) Semimembranosus. From R. M. Letts, D. A. Winter and A. O. Quanbury (1975), *Canadian Medical Association Journal*, **112**, 1091–1094.

force acting on it due to gravity, always acts vertically through a point called the centre of gravity of the body. So we can consider the mass of the body to be concentrated at that point, which is also called the centre of mass. That is, to simplify matters we can assume that the entire mass of the body is concentrated at the centre of gravity (centre of mass) of the body, a point which can be determined in a cadaver, for example, by hanging it up from several points and noting where the lines intersect which run directly down from the points of attachment. It can more readily be calculated approximately by cutting out a photograph of a species in various stances, as the Russian zoologist P. P. Gambaryan has done, and placing these on a Borelli stand (fig. 4.5). When the photograph on the Plexiglass sheet balanced, a line was drawn above the prism balancing edge; after it had been placed in several positions and had several lines drawn through it, their point of intersection was considered to be opposite the centre of gravity. In human beings this lies in the mid-line at a distance from the ground corresponding to about 55 per cent of the total stature (with individual variations

of ± 1.25 per cent), about 50 mm below the navel, and just anterior to
the vertebral column. If an animal's mass is considered to be concentrated at the centre of gravity, the legs themselves can be thought of as
massless levers belonging to the body. It becomes possible, in a somewhat idealized fashion, to consider the displacement pattern of the
centre of gravity during locomotion as constituting the summation or
end result of all the forces and motions involved with the movement of
the body from one point to another.

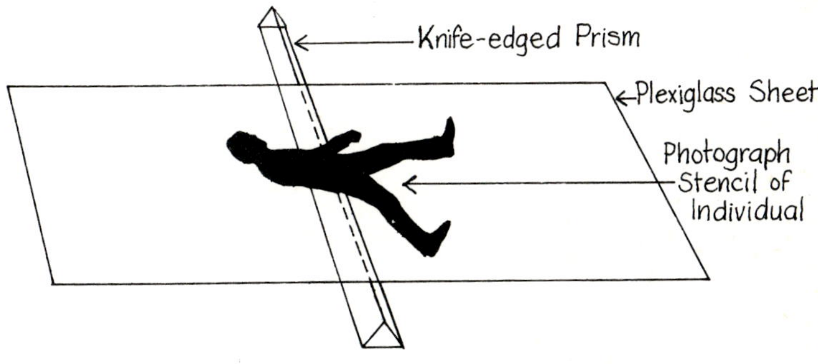

Fig. 4.5. Borelli stand used to locate the centre of gravity of a species.

4.5. *Walk—Definitions*

It is best to consider the human walk before that of four-legged
animals, because the movements of two legs are easier to understand
than those of twice that number. Fortunately, terms such as step and
stride can be applied to animals with any number of legs. A *step* is the
motion a single leg goes through from the time it is lifted off the ground
during locomotion until it is next lifted off again; it comprises a *stance*
or weight-bearing phase, in which the leg successively restrains forward
motion, produces balance and support, and finally exerts propulsion,
and a *swing* phase (fig. 4.2). A *stride* includes the movements of all legs
beginning at one leg and ending at the same phase of the same leg the
next time it is in that phase. *Cadence* is the rapidity with which steps
are taken. Since *speed = length of step × frequency of step*, it can be
increased by increasing either of these, or by increasing one of them
(usually frequency) with a less than proportionate reduction in the
other. An old person usually walks more slowly than a young one by
taking shorter and slower steps, with his toes pointed out. These features
of geriatric gait probably reflect a need for additional stability during
locomotion. Pregnant women often have an unusually slow gait

coupled with a tendency to lean back to counteract extra abdominal weight.

4.6. *Walk—Gross Displacements*

During walking there is up-and-down movement of the centre of gravity, less noticeable than in running, unless perhaps one person in a line of people is out of step, and bobbing conspicuously along. The total amount of vertical displacement in walking adult males is about 46 mm, the summit occurring in the middle of the stance phase of the supporting leg and the swing phase of the non-supporting leg, and the trough midway between these points when both feet are supporting the body. The maximum vertical distance from the ground in a walking person is always slightly less than his height when he is standing still, so that a man may walk through a tunnel that is exactly as high as his standing height without his head touching the ceiling. Indeed he will have over 10 mm clearance, although this is difficult to believe unless one tries the experiment.

As well as moving up and down as a person walks or runs, his centre of gravity also moves horizontally (fig. 4.6). If you walk along with your hands on your hips you can feel your weight shifting to the right as your right leg alone supports your body, and to the left as the left leg takes over the supporting function. The path of the centre of gravity is smooth, as it was in the vertical movement, but it follows a single- not a double-peaked curve during one stride. The magnitude of horizontal displacement in a walking adult is about 44 mm from extreme right to extreme left. As a walking person steps forward on his right foot, the

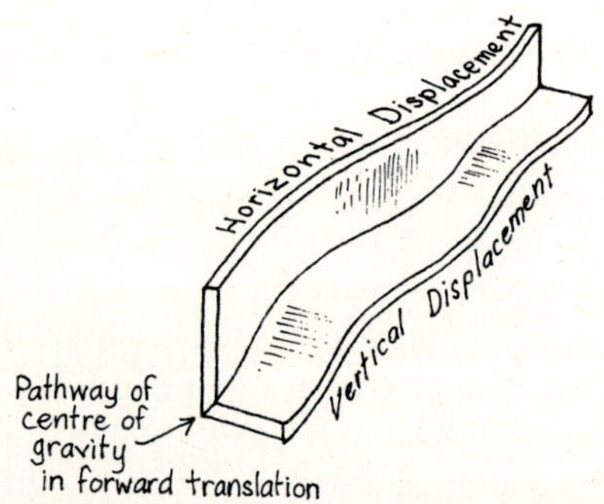

Fig. 4.6. Diagram of the path of the centre of gravity of a walking person. During one stride the vertical displacement undergoes two waves while the horizontal displacement undergoes one. Based on the work of J. B. D. M. Saunders, V. T. Inman, and H. D. Eberhart (1953), *Journal of Bone and Joint Surgery*, **35A**, 543–558.

centre of gravity moves upward and to the right, then down as the left leg completes its swing forward, and up and to the left as his weight is shifted to, and supported by, the left leg. The cycle is completed with the centre of gravity returning down and towards the right and beginning to move upward as the right heel again strikes the ground.

Because the centre of gravity is located inside a person's body, its changes cannot actually be measured during locomotion. However, vertical changes in the pelvis which can be monitored in a walking person are close to the vertical changes in the centre of gravity (fig. 4.7). Dr. V. T. Inman of California, who measured such changes, found that movements in the pelvis, hip and knee help to decrease the amplitude of the vertical motion of the centre of gravity, while those in the knee, ankle and foot are especially involved in smoothing out irregularities in the sinusoidal curve.

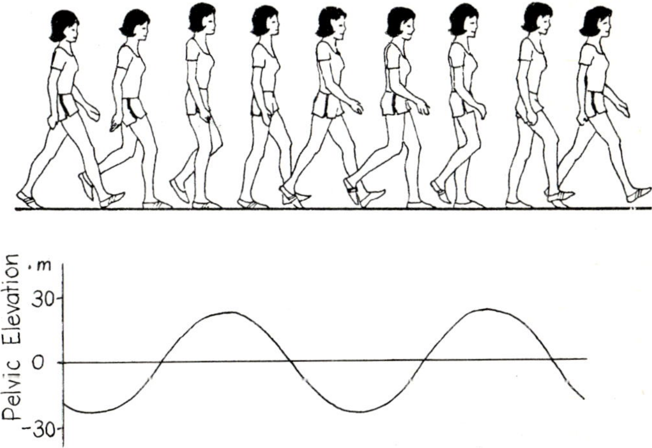

Fig. 4.7. Pelvic elevation during walking. The path of a point opposite the second sacral segment is graphed. During one stride the pelvis goes up and down twice. Based on the work of V. T. Inman (1966), *Canadian Medical Association Journal*, **94**, 1047–1054.

4.7. *Walk—Pelvic Movements*

Horizontal movements of the pelvis during walking are more complex than vertical movements, since they involve two components—rotation and tilt. The rotation during a walking stride averages about 6° to 8°, and is centred alternately about each weight-bearing hip joint (fig. 4.8). Thus the pelvis on the side of the swinging leg starts to rotate forward as the toe pushes off, and stops when the full weight is again placed upon the foot. These horizontal rotations lessen the amount of fall of the centre of gravity with each step, and extend through the leg itself as well as the pelvis. The arms swing out of phase with the pelvic rotation, apparently to dampen it.

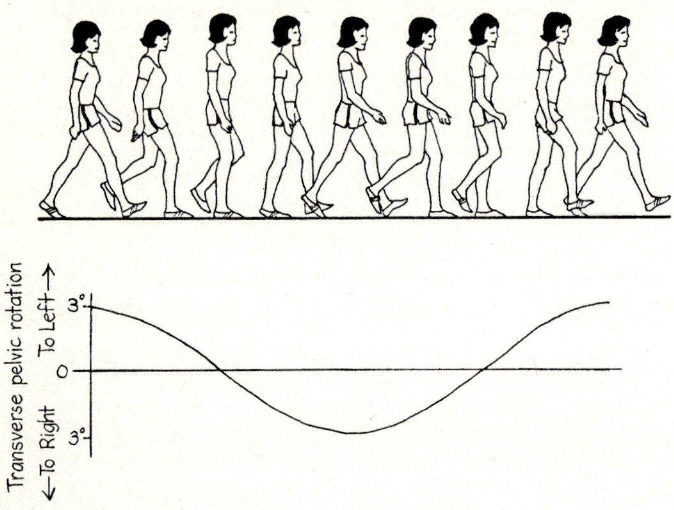

Fig. 4.8. Transverse pelvic rotation during the walk, as measured in a horizontal plane. During one stride the pelvis goes through one rotation. Based on the work of V. T. Inman (1966), *Canadian Medical Association Journal*, **94**, 1047–1054.

Pelvic tilt, which also averages 6° to 8°, can be best noted from the front or back view of a walking person. The pelvis tilts sharply down to the left as the left leg begins its forward swing, then gradually becomes level again by the time the left heel touches the ground at the end of the swing (fig. 4.9). This tilt tends to decrease the amount of time that the centre of gravity and the torso must be elevated at each step, thus conserving energy. To prevent the free leg from hitting the ground as it swings forward, lengthened as it is by the tilt of the pelvis downward to that side, the knee bends to about 110° midway through its swing phase (fig. 4.10).

4.8. *Walk—Force and Energy Changes*

So far the actual movements of the body during a walk have been considered; now we can look at the force and energy relationships which are developed in this gait. By means of force-plates it is possible to monitor the force which each foot exerts against the ground as one walks. From fig. 4.11 it is clear that the maximum vertical thrusts of a leg exist just after the heel is set down upon the ground, and just before the final push-off of the toe. The maximum horizontal force is exerted at the final toe thrust, as one would expect.

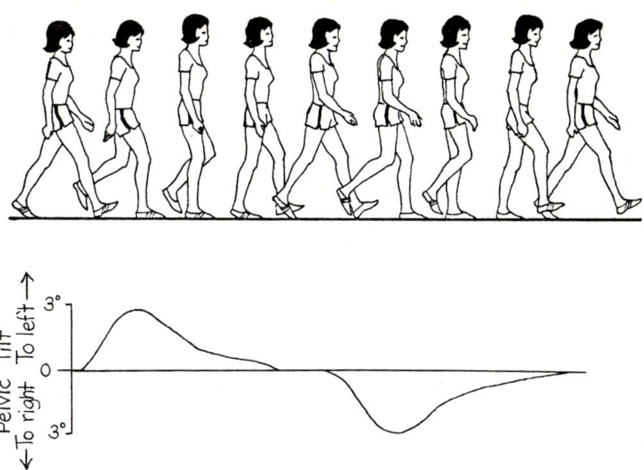

Fig. 4.9. Pelvic tilt during the walk. As the left heel touches the ground, the pelvis drops markedly to the right, non-weight bearing, side; it slowly recovers its position as the right leg swings forward. Based on the work of V. T. Inman (1966), *Canadian Medical Association Journal*, **94**, 1047–1054.

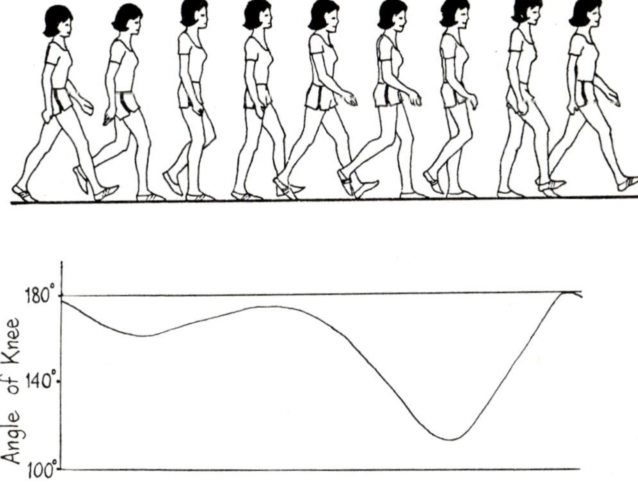

Fig. 4.10. Angular displacement of the knee during walking. The knee is more bent in the middle of the stance phase than it is at heel-on or toe-off. Based on the work of V. T. Inman (1966), *Canadian Medical Association Journal*, **94**, 1047–1054.

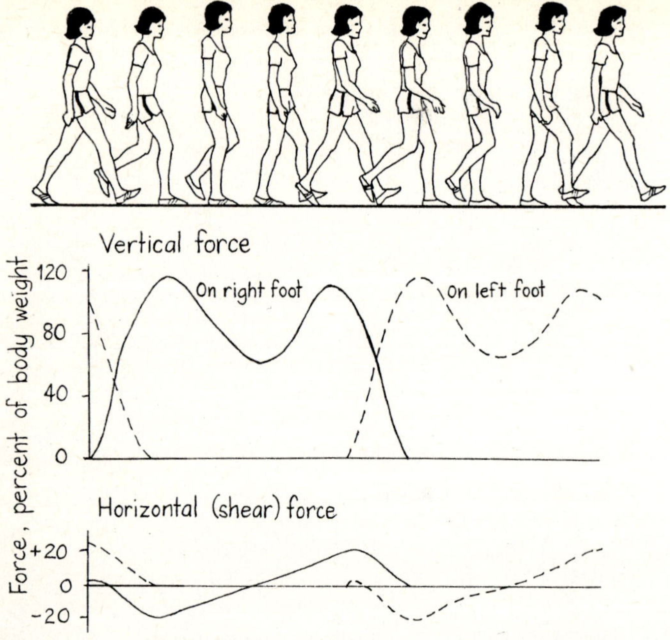

Fig. 4.11. Floor reactions during the walk as recorded from force-plates. The vertical forces are largest as the heel touches the ground and as the toe leaves it. The horizontal force is greatest as the toe pushes off the ground. Based on the work of V. T. Inman (1966), *Canadian Medical Association Journal*, **94**, 1047–1054.

As the centre of gravity of the walking body rises to its greatest height, when the supporting leg is fully extended, its potential energy is maximal (fig. 4.12). Some of this potential energy is converted into kinetic energy as the body rolls slightly forward about the stationary foot as axis and the centre of gravity falls. From the energy point of view, the body behaves like an inverted pendulum, of which the pivot is made to move forward in steps by the legs. With the next elevation of the body, the forward velocity decreases, with some of the kinetic energy being re-converted into potential energy. Of course, work is also done by the leg muscles which convert *internal* energy into mechanical work. These provide the thrust of the stationary leg before it moves forward, and initiate the hip and knee flexions of the swinging leg, which gains considerable kinetic energy as it swings forward; meanwhile the leg itself is decelerated through the action of the hamstring muscles. The cyclic pendulum-like transfer of energy within the body from potential to kinetic minimizes the work the muscles must do in walking, work estimated by recalling how tiring walking uphill can be, and how

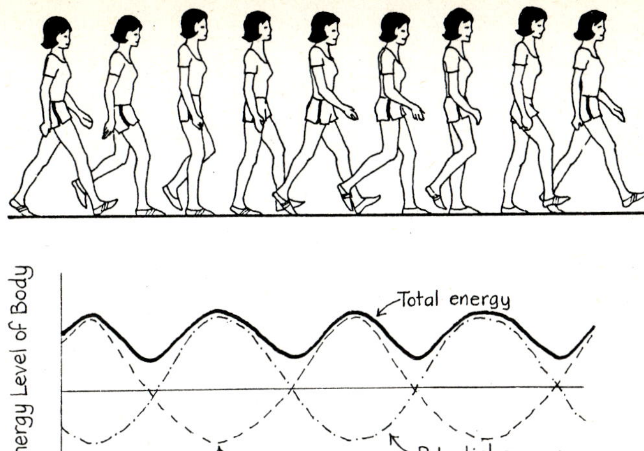

Fig. 4.12. Idealized curves showing the transfers of energy within the body during the walk. If the system were conservative, the total energy curve would be a straight line. The energy contributed by the musculature of the body is indicated by the difference between the total energy curve and a straight line. Based on the work of V. T. Inman (1966), *Canadian Medical Association Journal*, **94**, 1047–1054.

pleasant walking downhill. A descending gradient of about 1 in 25 allows one to proceed without doing any work in raising the centre of gravity of the body.

4.9. *Walk—Muscle Action*

So far little has been said about muscle action, the basis of all movement. A relatively simple example may be considered in the upper leg, where the hamstring muscles at the back of the thigh originate on the ischium and insert on the posterior surfaces of the tibia and fibula, causing the leg to flex at the knee and/or extend at the hip, while the antagonistic quadriceps muscle at the front of the thigh originates on the ilium and inserts on the anterior surface of the tibia (the tendon incorporating the knee cap as it passes over the knee joint), causing the leg to extend at the knee and/or flex at the hip. When electromyographs were made of these muscles during walking, it was found that the hamstring muscles reached a peak of activity in extending the hip joint just as the heel struck the ground. The quadriceps muscle contracted slightly after the hamstrings, acting to extend and stabilize the knee as the body swung forward over the leg. This muscle contracted again at toe-off, this time bending the hip as the leg swung forward. In some people the hamstrings may also be active at toe-off, apparently to bend the knee as the leg begins swinging forward. The action of a leg muscle depends on whether the leg is swinging free or set on the ground. Because a muscle produces movement only by contraction, which draws the points of attachment of the muscle closer together, the bone at the end of the muscle which is the lighter is the one which will move. In general, if the hamstrings contract when the leg is swinging free, they will bend the knee, but if they contract when the foot is on the ground, they will extend the hip (fig. 4.13). The actions of the quadriceps are the reverse of these.

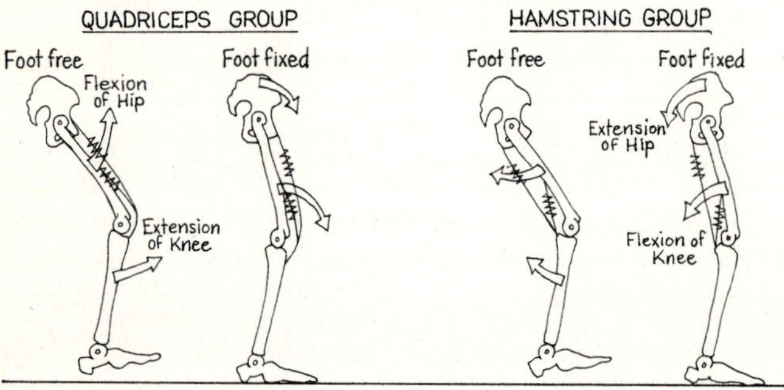

Fig. 4.13. Actions of the upper leg muscles. The line of action, noted by the arrows, depends on whether the foot is on the ground or is swinging free. Based on the work of C. W. Radcliffe (1962), *Artificial Limbs*, **6**, 16–24.

4.10. *Walk—Variations*

We have already mentioned that old people and pregnant women have abnormal ways of walking. It is surprising to find then that such people, and people with other physical differences, often have walking patterns similar to the normal. This is because people, like other animals, have a striking capacity to compensate for abnormality. The walk of a woman in high heels has a number of unusual components, yet force-plate records show that her vertical accelerations and the path of her centre of gravity are similar to those of a woman in low heels. Because in high heels the foot cannot move naturally, the movements of the ankles, knees, and hips are modified to counteract this deficiency. The initial knee flexion is increased, the pelvic tilt and rotation are exaggerated, and the step is small, which gives the woman a characteristic but not ineffective gait.

Similarly if a man has a stiff knee he can compensate fairly well for this condition with his other leg joints (especially if the knee is fixed at about 15–20°), maintaining a relatively normal path of his centre of gravity. However, a great deal of energy is needed to initiate the swing phase on the side of the stiff knee. Early flexion of the knee at the start of the swing phase is important, since it converts the leg into a double pendulum and thus decreases the angular momentum of the whole lower extremity. If the knee cannot be bent, the hip flexors alone must initiate the swing phase, demanding that a force be exerted by the hip flexors which is several times as great as normal. If an old person loses the function of two joints, physicians have found that his life is likely to be shortened because of the strain on his cardiovascular system, which must supply locomotion requirements at a rate of up to 300 per cent of normal.

4.11. *Walk—Medical Research*

Only recently has research been instigated to help people with abnormal gaits to move more easily. A simple method of assessing pathological gaits is to record the stance and swing phases of the walk, as has been done for various people including one with a painful hip, one with a partly paralysed leg, and one with Parkinson's disease (paralysis agitans). The first person (fig. 4.14(B)) walked much more

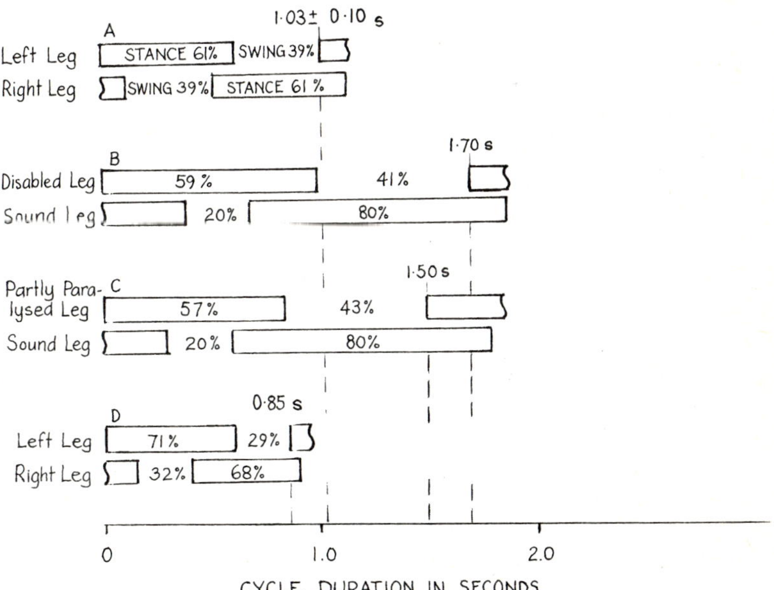

Fig. 4.14. Duration of stance and swing phases in four walks. A. Normal men; B. Patient with a painful hip; C. Patient with a partly paralysed leg; D. Patient with Parkinson's disease. Based on the work of M. P. Murray (1967), *American Journal of Physical Medicine*, **46**, 290–333.

slowly than a normal man (fig. 4.14(A)), spending less time being supported by the disabled leg than by the sound one; any asymmetry of human gait between right and left legs indicates an abnormality of some sort. The patient with the partly paralysed leg (fig. 4.14(C)) had the same pattern of support of his legs as the first, with his sound leg doing most of the work, but because movement was not as painful to him he walked more quickly. Because people with Parkinson's disease suffer from involuntary trembling and a tendency for their bodies to bend forward at the waist, they take short, quick steps to keep themselves from losing their balance. This is evident in the diagram for a patient affected with the disease (fig. 4.14(D)).

A more comprehensive analysis of a pathological gait can be made by film plus electromyographic (EMG) studies. For example, R. M. Letts, D. A. Winter, and A. O. Quanbury in Winnipeg, Canada, analysed the gait of a spastic 13-year-old boy who had his knees bent at an angle of 35° and who walked very slowly, with a stride of only 125 mm. His left leg rotated in at the hip as it swung forward, while the hip bent to an angle of 40°. The EMG explained these movements: the left rectus femoris (part of the quadriceps) was hyperactive through the walking cycle, while both left hip adductors and medial hamstrings showed bursts of activity at the beginning of the swing phase which tended to produce the excessive internal rotation at the hip joint (fig. 4.15(A)). As a result of this assessment of muscle activity, an operation was performed in which the left rectus femoris was lengthened so that the knee was less bent, and the left hamstrings and adductors were lengthened to curtail the excessive rotation and allow the boy a longer stride. His gait was much improved after the operation, with his stride length doubled, his knees less bent, and the rotation much reduced. The EMG showed that the rectus femoris activity was much quieter and the adductors were, electromyographically, almost silent (fig. 4.15(B)).

4.12 *Running*

Running in man differs from walking in that the support of the body is not shared by both legs at any one time, but is borne first on one leg and then on the other. Between periods of support the body moves freely through the air (fig. 4.16).

The up-and-down movement of the body is especially evident, with the head highest when the feet are not on the ground, and lowest when one foot, with knee bent, supports the body. Thus there are two peaks and two troughs in the smooth movement of the centre of gravity during one stride. The thrust for the movement is supplied in turn at each supporting foot: the upward component has a constant value,

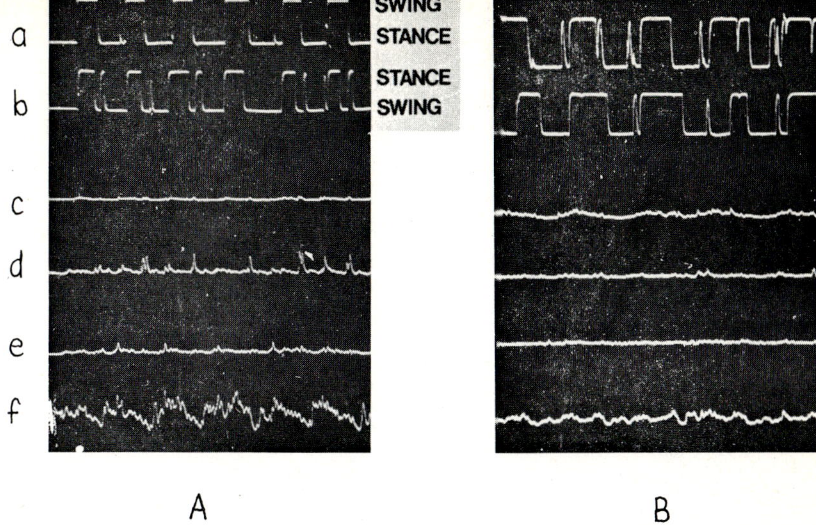

Fig. 4.15. EMG signals from the leg muscles of a boy with spastic diplegia. A. Preoperative signals; B. Postoperative signals. (*a*) Stance and swing phase of right leg; (*b*) Stance and swing phase of left leg; (*c*) Left gluteus maximus; (*d*) Left adductors; (*e*) Left medial hamstrings; (*f*) Left rectus femoris. From R. M. Letts, D. A. Winter, and A. O. Quanbury (1975), *Canadian Medical Association Journal*, **112**, 1091–1094.

Fig. 4.16. Running. Runner pushes off from the ball of the foot but is supported by the entire foot. Arms move opposite to the legs.

independent of speed, of about the same magnitude as the body weight; the horizontal component is directly related to speed and also supplied by the forward component of the reaction of the ground to the push of the foot against the ground.

In walking, the two curves of kinetic and of potential energies of the body during a step cycle are almost directly out of phase, as we have seen (fig. 4.12). Some of the kinetic energy of a subject is turned into potential energy during the lifting of the body in the first phase of the step, while during the second phase this potential energy is turned back into kinetic energy. During slow walking nearly all the kinetic energy comes from the potential energy acquired during the first phase of each step when the body is lifted. In faster walking the push is no longer almost entirely vertical, the component which increases the potential energy, but somewhat horizontal as well, a trend which increases noticeably in running. Because in running potential and kinetic energy are not interchanged to any extent, the work done against gravity and the energy needed to sustain the velocity-changes in the forward direction are both provided by muscular internal energy and take place simultaneously. In both walking and running the work done increases the faster a person goes, more so because of forward velocity-changes than because of changes in the height of the centre of gravity. As one walks faster than about 7 km h^{-1}, the maximal value of kinetic energy is reached progressively earlier than the minimal value of potential energy, and the work done by the foot as it pushes off is progressively greater. When the vertical component of such push reaches a value higher than the body weight, the body loses contact with the ground; the two curves of potential energy and of kinetic energy are in phase. At this point, walking shifts to running. The external work per kilometre done by a person while walking at 12 km h^{-1} is similar to that done while running at the same speed, but the overall energy expenditure as measured from the oxygen consumption in subjects moving on a treadmill is greater for walking than running. The internal energy used up in fast walking is greater than in running, in part because of the greater stress placed on the supporting legs.

Although a person running at constant speed on flat ground uses up a great deal of energy, which is needed both to overcome frictional and viscous resistance in the joints and muscles and to produce the continuous accelerations and decelerations of the mass of the body and of the limbs, he is doing no external work in the physical sense of *"work = force × distance"*. If he runs up a hill, however, he expends internal energy and does do physical work as he has gained potential energy (the product of his weight times the vertical distance the centre of gravity has risen). If he runs downhill, he has lost potential energy and gained kinetic energy—we can say he has done "negative work", a term used to denote that external work has been done on him, not by him. It has been calculated that during level running a mouse weighing 30 g uses about eight times as much energy per unit of body mass as does a girl weighing 18 kg, but when these two run uphill, they use the same relative amount of energy, as the mechanical work involved in lifting one kilogram one metre vertically is the same for both. Thus small animals can be very energetic in climbing, using relatively little more energy than that used in level running, and rush up trees at about the same speed as that with which they bound along the ground, a feat impossible for large animals. When any animal runs downhill, it does so readily, losing on the way down most of the potential energy which had been gained during uphill running.

4.13. *Locomotion on the Moon*

Recently, with the prospect of men on the Moon, the Italians R. Margaria and G. A. Cavagna analysed the potentialities of human locomotion there. The body mass m of a person will be the same on the Moon as on the Earth. The weight W of the person will be different however, because the force of gravity exerted by the Earth on a person at the Earth's surface (which is what we mean by weight) is much greater than the force of gravity exerted by the Moon on the same person at the Moon's surface. The effect of gravity can be specified in terms of the intensity of gravity g (in newtons per kilogram), more often quoted in the equivalent form as the free-fall acceleration g in metres per second per second. The approximate value at the Earth's surface is $g_E = 9{\cdot}8 \text{ N kg}^{-1} \equiv 9{\cdot}8 \text{ m s}^{-2}$, and at the Moon's surface $g_M = 1{\cdot}6 \text{ N kg}^{-1} \equiv 1{\cdot}6 \text{ m s}^{-2}$. Thus for a given mass m, the weights are

$$W_E = mg_E$$

on the Earth and $\hspace{10em}$ (1)

$$W_M = mg_M$$

on the Moon. However, the mass, and hence the force required to produce a given acceleration, does not depend on gravity. If F represents the force acting on one from the ground as one pushes off with the foot at each step and a is the acceleration in the direction of progression, then

$$F = ma. \hspace{6em} (2)$$

Thus the ratio of the weight W to the accelerating force F is the same as the ratio of g to a. This follows from equations (1) and (2) which give

$$\frac{W_E}{F} = \frac{g_E}{a} \text{ on the Earth}$$

and

$$\frac{W_M}{F} = \frac{g_M}{a} \text{ on the Moon.}$$

This change in the relative importance of the weight-force W acting on the body and the accelerating force F is one of the main factors responsible for the change of mechanics of locomotion where the intensity of gravity is less than it is on the Earth.

Since on the Moon the gravitational force is about one sixth that on Earth, a person there would experience a weight-force only one sixth of what he experiences on the Earth. Because of this reduction the potential energy gained in the first phase of each walking step would also be correspondingly less, as would the subsequent kinetic energy into which it was transformed. Thus there would be little energy available for forward progression on a horizontal surface. The falling forward of the body in the second phase of each step, under a greatly reduced gravitational force, would be very slow. One may calculate that if 100 steps per minute is a normal walking rate on Earth, this rate would be reduced to 40 steps per minute on the Moon, although with relatively little expenditure of energy.

Running could also be frustrating on the Moon, because (with flat unroughened surfaces) the maximum (limiting) frictional force between a person's foot and the ground depends on his weight. Thus the maximum possible acceleration, unless spikes or crampons were used, would be much less on the Moon than on Earth since the maximum possible push on a foot for a given kind of surface is only one sixth as great. The length of each step will be similar on the Earth and on the Moon, but according to calculations the step frequency on the Moon will be decreased to a maximum of about 160 per minute on hard soil, or about 100 per minute on ground where the friction is low. On the Earth a top sprinter has a step frequency of about 250 steps per minute.

In general, speed of running depends on the step frequency, the force of propulsion, and the inclination of the push of the foot on the ground (at a minimum of 45°). On the Moon, a push with the vertical component appreciably greater than the body weight can be given, because the muscular force of the body is the same as on the Earth, while its weight is reduced to one sixth. Thus a person can more readily leap into the air at each step in a succession of jumps than he can run. However as jumps take longer to negotiate on the Moon because of the reduced gravitational force, the number of steps is less than it would be in running. Even so, with powerful jumps the horizontal component of each can give a person considerable forward momentum. It seems possible that one could progress in jumps on the Moon as quickly as one could run on the Earth.

When American space researchers working for the American Government actually carried out locomotion experiments in a simulator set to approximate to gravity on the Moon, they found that a man could walk at a speed of 5 km h^{-1}, far faster than had been calculated theoretically by Margaria and Cavagna. This was because, far from walking upright as it had been supposed a person would do, the men being tested found this an uncomfortable and even an unnatural position. Instead they leaned forward as they walked, some so far that they fell on their faces. With their bodies inclined forward, the acceleration phase within each step of their walk increased, and the deceleration phase decreased. Swimmers walking under water lean forward in the same way because they are buoyed up by the water.

The above calculations have not taken into account the limitations on movement set by the pressure space suits which astronauts must wear. The suits worn in the American Moon missions did allow the astronauts to jump readily and to walk slowly, as anyone who watched their televised activities will have seen.

5. Athletics

5.1. *Sprinting*

Foot races at Olympia in ancient Greece were of two types, as they are today: sprints and long-distance races. Each demanded special skills. Apparently Lados of Achaea, who crossed the finishing line of a 370-metre race many metres ahead of his competitors, was so elated by his success that he ran all the way to his home, many kilometres away, to bring the good news to his family and friends. However, because he was primarily a sprinter, not a long-distance runner, the cross-country run was too much for him, and he died shortly after reaching his home. On the other hand Drumos of Epidaurus, a famed long-distance runner, ran directly home after winning the 4·8-kilometre race at the Olympic Stadium, because he also wanted to be the first to proclaim his victory to his city. He covered the 144-kilometre trip easily because he was trained to run such a distance.

Early drawings of runners on Greek vases show that their styles were similar to those of modern athletes. Sprinters ran with arms held high and legs pushing strongly off from the balls of the feet (fig. 5.1). Today sprinters are coached in a similar high knee action, with long strides, a negligible kick-up of the legs behind, and energetic (but less high) arm movements. Many coaches believe that the faster the arms move, the faster the legs will go. Sprinters then, as now, leaned forward at an angle of about 25° to the vertical. No coaching however can make a fast sprinter out of a slow one; sprinters are born, not made.

Years ago Wallace Fenn, a student of athletics, compared the running style of sprinters to determine why some were fast and others slow. The fastest runner was outstanding in that his step was long (up to 1·94 m, while most of the other 33 men had a step less than 1·85 m), and his cadence fast (4·35 steps per second, compared to most with fewer than 4 steps per second). In style the fastest runner lifted his thighs much higher than did the other runners; his thigh at its highest was 7° below the horizontal, while most of the others lifted their thighs only to about 26° to the horizontal.

Recently, in 1973, Dr. Brandell of the University of Saskatchewan

Fig. 5.1. Sprinter copied from a Greek vase. There is high leg and arm action.

in Canada compared the body movements of a 100-metre sprinter with those of a man running 1500 metres. In the sprint the stance phase only lasted 19 per cent of the time of a stride, whereas in the long-distance run it lasted 28 per cent. In the latter the stride lasted longer (0·72 s instead of 0·59 s), but was of shorter length (4·4 m instead of 5·3 m), with greater amplitude (but slower velocity) of the knee joint (42° instead of 26°), ankle joint (72° instead of 64°), and thigh motions (58° instead of 50°). In the sprint therefore the limbs go through relatively shorter but quicker angular motions which provide great thrust, while spending relatively more time in the unsupported "float" phase and less in the support phase. An important difference between the sprint and the distance run was that in the latter the foot first touched the ground with the heel, which caused a slight reversal in the forward direction of the thigh and a knee flexion of about 26°. In the sprint the ball of the foot touched down first, so that the forward motion of the thigh was only slowed but not reversed, and the knee joint only flexed 10°. In the sprint a more continuous forward thigh motion was achieved by keeping the body weight on the ball of the foot, and absorbing the shock of contact at the ankle joint by the calf muscles, rather than at the knee joint by the quadriceps femoris muscle. Usually good sprinters are of medium height, muscular, relatively short-legged, and with a slender body build. The 400-metre runner tends to have longer legs, and the distance runner similar proportions but a slighter build.

5.2. *Long-distance Running*

Early Greek long-distance runners, like modern ones, had an entirely different posture from sprinters as they jogged along in a more erect and relaxed fashion (fig. 5.2). Their arms were held near the trunk and their entire foot rather than the ball alone supported and propelled them forward in a ball–heel–ball motion. Good long-distance runners may vary considerably in style, some using relatively short strides and others a high kick-up at the back. All hold their bodies leaning forward at about a 10° angle to perpendicular. Greek runners may have been stronger than modern ones. One early Greek race which is unknown today is the 370-metre armour race in which men ran naked, but carrying a heavy bronze shield and wearing a large bronze helmet and sometimes metal greaves on the legs. In all the armour weighed over 20 kilogram.

At the start of a race at Olympia, the contestants lined up at a slab of stone set into the ground. This stone (it is still there) has two parallel

Fig. 5.2. Long-distance runner copied from a Greek vase. The man is supported by his whole foot, with erect posture and relaxed arm movements.

grooves perpendicular to the track cut into it 180 mm apart, in each of which the runner curled the toes of one foot. He began a race standing up, but with his trunk bent forward and his arms thrust ahead of him. As the starter sounded he could push forward with some force, but much less than athletes of today gain with the crouch start. This start was first used competitively in 1888 by Charles Sherrill of Yale University. When the starter noticed Sherrill bent over, he delayed the race until he was sure Sherrill felt well and knew what he was doing. A reporter noted that although Sherrill appeared to stumble at the start, he recovered himself and went on to win the race. Since then the superiority of the crouch over the standing start has been so obvious that no other method is used.

No matter how fast a normally shod person runs, his foot never pushes against the ground at an angle less than 45° to the ground. A smaller angle would mean that the horizontal force would be greater than the weight of the runner, and even if he could keep this up, the runner's foot would almost certainly slip. For the crouch start therefore runners use blocks which give their feet a solid, sharply angled surface against which to push. Some athletes place their starting blocks close to the starting line so that just before the gun their weight is largely on their fingers. Others place the blocks farther back so that their weight is largely over their legs until the last few seconds; at the gun the angle of forward drive is low and the legs can take longer strides than is usual. During running, although the foot at each stride leaves the ground at an angle greater than 45°, track shoes with nails or studs are still an asset for runners to prevent slipping.

5.3. *Hurdling*

A good hurdler considers hurdling a fast sprint incorporating a few obstacles, rather than a series of jumps that must be negotiated. The hurdles should be mastered in such a way that a person watching a hurdler's head does not know exactly when each hurdle is being jumped; this means that there is little change in the height of his centre of gravity beyond that encountered in sprinting. It also means that a hurdler must bend his body close over his front leg as he goes over each obstacle. There is no such thing as too much forward lean in hurdling. Speed is essential, not only to ensure a good performance, but to enable the athlete to reach each obstacle in the few steps allowed. The take-off distance should be two metres or more ahead of each obstacle so that the hurdler's momentum will carry him over it, without his consciously having to jump it. The lead leg, which is brought up straight ahead of the hurdler, begins to cut down even before it has crossed over the obstacle. The cut-down distance must be as short as possible to save time and prepare for the sprint to the next obstacle (fig. 5.3). An athlete can gather speed only when his feet are touching the track, so time spent floating in the air is wasted. As he passes over each obstacle at least one of a hurdler's arms is stretched forward, to help keep his weight forward. The trail leg is brought up at the side of the hurdler's body, so that the thigh, lower leg, and foot are parallel to the top of the hurdle. This leg should be driven to the ground as quickly as possible after it has crossed the obstacle to maintain momentum.

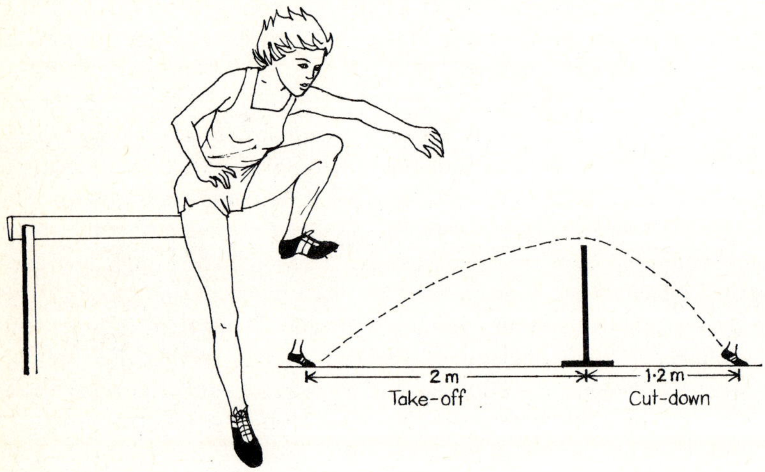

Fig. 5.3. Hurdling. The leading leg is brought down as soon as possible after crossing the obstacle to save time.

5.4. *Long Jump*

Long jumping was as popular at Olympia as running, with athletes completing tremendous jumps, probably greater than modern records, not necessarily because they were better athletes but because they carried two metal or stone weights called *halteres*, one in each hand. These, weighing from 1 to 4·5 kg, could add considerable distance to a jump if used properly. The athlete gave them linear momentum with his arms just before taking off using his legs, so that they effectively pulled him forward; they were then used to change the body's balance while in flight. A contestant could use *halteres* of any weight, shape or size; those in museums today have a hole in them for grasping, or are shaped with a handle for holding plus an inflated end for added weight. Probably each man practised with *halteres* of a variety of shapes and weight to determine which best suited his personal style.

A 4·5 kg mass could increase an athlete's momentum considerably, but he must be strong enough to control it. For the running long jump a contestant ran five metres, took off at the starting line, flung his weighted hands forward as he launched into the air (fig. 5.4(*a*)), and swung them behind him just before he landed to give himself a few more centimetres, effectively swinging his feet that much ahead of the changed centre of mass (fig. 5.4(*b*)). As he landed on flat feet he brought the *halteres* forward again to his trunk, to steady himself. The jump was judged on grace and style, as well as on distance.

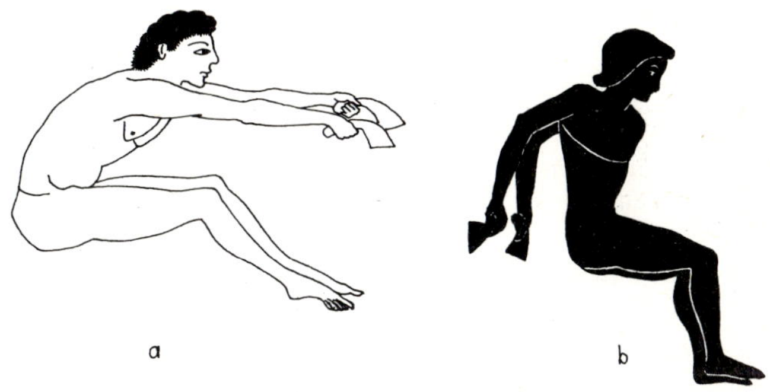

Fig. 5.4. Long jump using *halteres*. Olympic figures copied from Greek vessels (*a*) in the middle of a long jump and (*b*) just before landing.

For the standing long jump an athlete stood on flat feet, swinging his *halteres* back and forth until they had a large amplitude; then as his arms swung forward he leapt, thrusting with his legs, so that the momentum of the *halteres* was added to the momentum of his leap. In the air he snapped his legs up parallel to the ground and to his outstretched arms, landing as for the running jump.

Today coaches of the long jump emphasize the need for maximum speed at the run up, with maximum height following the take-off. Ideally, just as in the high jump, a jumper should have his centre of gravity directly above his take-off foot at the moment of take-off. He should stamp down hard with this foot, then straighten this leg as quickly as possible in the spring upward. His head should look up and his arms stretch upward too. The style used in the suspended part of a long jump is variable; some champions make walking movements in the air, some remain still, some arch the back at first, before drawing the legs and arms forward for the landing. The feet should touch ground ahead of the bent body, but the upper body must have enough forward momentum so that it is then thrust forward over the bent knees, with the buttocks just clearing the sand. The measure of the jump is taken from the farthest back point of landing, which should be the heels.

5.5. *High Jump*

Until 1929, the rules on high jumping stated that one foot of a jumper must cross the bar before his head; he could not dive or somersault over the bar. Now this rule has been replaced so that any style of jump is acceptable, unless it includes specially designed shoes. Just before 1960, the Soviet Union's high jumpers were having great success using shoes that were up to 2·5 cm higher on the sole than on the heel. At the instant of take-off, a jumper was able to rock far up on to his toes. This extra rocking movement was such a great aid in gaining height that a rule was quickly passed limiting the thickness of a jumper's shoes.

The important effect of the height of the centre of gravity can be emphasized by considering a person high jumping. The higher he raises his centre of gravity during a jump, the more energy he must use. The most effective jump will therefore be one in which his centre of gravity is raised only just high enough to clear the bar. In the old-fashioned scissors high jump the centre of gravity had to be raised nearly 30 cm above the bar for success, whereas in this jump combined with a lay-out, in which the back goes over the bar horizontally and nearly touching it, it is only raised about 18 cm. In the Western roll (fig. 5.5(*a*)), in which the jumper goes over the bar sideways, with his legs held close together, the centre of gravity must be raised 15 cm above the bar. A human body is thicker from side to side than it is from front to back, so the straddle jump (fig. 5.5(*b*)), in which the body straddles the bar face down, with one leg on one side of the bar and one on the other, is more efficient even than the Western roll. By the same token the

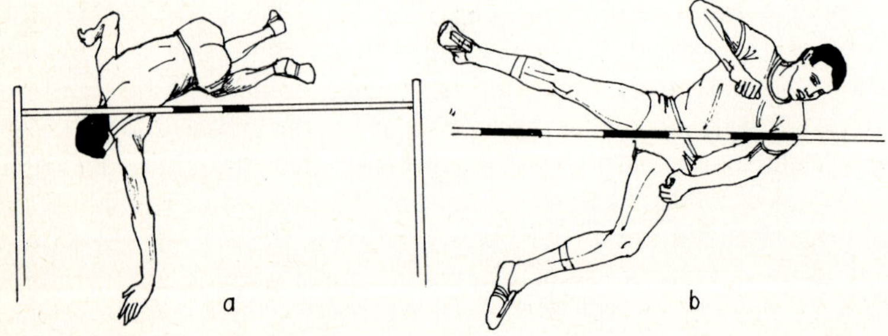

Fig. 5.5. High jumping. A person's centre of gravity must be raised higher (*a*) in the Western roll than (*b*) in the straddle jump.

forward-dive roll, in which a jumper dives over head first and twists past the bar to land on his back, has great potential. The high jump is an exciting event in that so many styles are possible, even to the relative speed of the run-up. For success however an athlete must have some native "spring ability". This can be measured by having an athlete mark with chalk a point on the wall as high as he can reach while standing, and as high as he can reach when jumping. If the difference between the marks is over 60 cm, the athlete has good potential as a high jumper.

5.6. *Racial Differences*

Interesting theories about racial differences in athletic contests have recently been postulated because it has been noted that mongoloid people, notably the Japanese, are especially proficient in international gymnastics, swimming, and wrestling; that Caucasians have dominated rowing, fencing, and sailing events; and that negroes have excelled in sprints, jumps, hurdles, boxing, and basketball. In the 1968 Olympics in Mexico City the eight finalists in the 100-metre sprint were all black. Some of the differences are cultural but some seem to have a physical origin. Negroes, for example, on the average have heavier bones, less body fat, and greater musculature than whites. Negroes tend to have longer legs and arms too, which help them achieve a longer stride in running and a greater height in jumping. Their hips are narrow, which gives them less angular reaction in each running stride, and their reflexes are exceptionally fast. The chest construction of the black could be a handicap however. Compared to an average white, his chest is shallower, and the slant of his ribs gives him less expansion of the chest wall in deep breathing. This relatively low vital capacity could explain why negroes are superb in explosive sports such as sprints and basketball, but less prominent in long distance races that call for unusual amounts of stamina.

6. Walking and Running in Four-Legged Mammals

6.1. *General Principles*

Complicated as human locomotion may seem superficially, there are four general principles which underlie it, and which can be stated in a series of pairs:

(*a*) *Trunk* versus *Legs*. During normal locomotion the body moves forward more or less at a constant rate, whereas the legs move in jerks; first the foot remains planted while the hip moves forward, and then the hip remains relatively still while the foot swings forward.

(*b*) *Horizontal* versus *Vertical Displacement*. Locomotion is never smooth. There are always up-and-down movements with each step, because the body is carried above the supporting leg alternately when it is erect and when it is at an angle; and sideways movements because the body is supported first on the right leg and then on the left.

(*c*) *Deceleration* versus *Acceleration*. Each leg as it is set on the ground acts to slow down the forward momentum of the body. Only after it has reached an angle of over 90° with the ground does its push begin to accelerate the body again.

(*d*) *Stance Phase* versus *Swing Phase*. Each leg spends part of each stride supporting the body (= retraction) and part swinging forward (= protraction). The supporting phase becomes shorter the faster an animal moves.

The above principles apply not only to human beings, but to all bipedal animals which move using alternate legs and to all non-jumping quadrupeds as well. For such quadrupeds it is possible to add several more general principles:

(*e*) *Hind Quarters* versus *Forequarters*. In most quadrupeds the force of propulsion for movement tends to come from the hind quarters while the forequarters tend to support the animal and, at fast speeds, to absorb the shock of landing following a period of suspension. The higher the speed, the greater is the propulsion.

(*f*) *Back Legs* versus *Front Legs*. At slow speeds a back and a front leg are used alternately for support; at fast speeds, in which the animal would be slowed down if it were supported by a hind leg after being supported by a front leg, the two hind legs are each used before the two front legs.

(*g*) *Lateral Legs* versus *Diagonal Legs*. Diagonal legs (left hind and right front, or right hind and left front) are often used as supports where stability is important. Lateral legs (both left or both right) are invariably used for support when the opposite hind foot is being placed on the spot where the front foot had been planted. This point and the others will be elaborated in this chapter.

(*h*) *Symmetrical Gaits* versus *Asymmetrical Gaits*. The symmetrical gaits are the walk, trot, pace, and bound; the asymmetrical gaits are those in which the right legs do *not* do the same actions as the left, primarily the gallop. Care must be taken not to confuse this generally accepted definition of an asymmetrical gait with that of P. Gambaryan who defines such a gait as one in which the forelimbs move first, followed by the hind limbs. There is thus no symmetry in the cycle, because the movement of the limbs in the first half is never repeated by symmetrically arranged limbs in the second half.

(*i*) *Principles* versus *Exceptions*. Beware of principles such as those above. Almost as soon as one is formulated, an exception arises. After I had written recently the final draft of a review paper on gaits, stating that elephants were too heavy ever to be supported by only one leg during walking or running, Gambaryan sent me a copy of his excellent book on mammalian locomotion which included pictures of an elephant walking fast, balancing on each single leg in turn. Fortunately I was able to change my statement before publication.

The major difference in studying the locomotion of human and non-human animals is that while people are usually willing to co-operate in studies and do what you tell them, other animals are not. Even domestic animals can be difficult to organize. Ten years ago I decided to undertake a comparative study of gaits in dogs, which were known to be able both to pace and trot; because the breeds come in a variety of shapes and sizes the characteristics could probably be related to differences in gaits. Unfortunately, what seemed a simple undertaking had incredible problems. At first I tried to work with a local Kennel Club, contacting members in turn and arranging to visit each to film his or her purebred animals. Many of the breeds recognized by the American Kennel Club have specifications for movement, gait, or action which must be met in judging, so I felt my studies would be of interest to dog breeders. The members themselves were most friendly, but they could not understand that I had to film their dogs *off* their leashes and *outside* their often cramped pens. On one occasion I drove 500 km to photograph a rare breed of dog, only to find that its owner refused to let it walk or trot anywhere but at her heel. Since she walked slowly, I had no chance to obtain footage of it moving at a more normal speed. Later I attended a Dog Show to hand out questionnaires dealing with dog movements, but none of these was returned. Not one handler was interested in whether his dog could pace as well as trot. When I discussed this question with a few off-duty owners, they were more communicative, but not helpful. One man explained that his dog often paced, usually back and forth across his bedroom window when he went out to get the mail. He described the dog's activity correctly, but he had no idea I used the word pace to define a particular way in which the dog used its legs, even though I tried to explain this. Nor were my professional colleagues more useful. One man kindly agreed to film his sporting dogs in full gallop, but because he was used to shooting he filmed the dogs in the same way as he shot, aiming the camera just in front of the target. The film showed many sequences of his dogs' noses, but none of their legs. Before I had accumulated enough film to do more than begin my study on dogs, a similar work by M. Hildebrand of California was published. My original idea had been a good one anyway!

No matter how carefully material on tame animals is collected, there is always the danger that there are undetected defects in the locomotion of pets which will mar the study. If possible it is best to use wild animals in the field, animals which are normal enough to survive the daily exigencies of obtaining food and avoiding predation. For this reason I undertook to study the gaits of wild giraffe in Africa rather than those in zoos, although film on the latter would have been much easier to obtain. In South Africa it was easy enough to approach giraffe, because they are now seldom hunted, but it was difficult to film sequences of their walking or running which were not obscured by long grass or bushes. Once I was discomfited by a cobra which approached me partly reared up, frightened by a giraffe's rush near it. I found that giraffe could be very lethargic; when I wanted to film a standing animal starting to walk forward, I wasted many reels because although various individuals *seemed* about to begin moving, they often remained still instead.

Because so many animals live where the terrain is rough or the vegetation dense, I was pleased recently to be able to work on the locomotion of the camel, which lives in the Sahara where the ground is almost perfectly flat and almost without vegetation. I was able to film camels to my heart's content—young and adults; males and females; camels loaded, unloaded, and pulling water at the wells; camels walking, pacing, and (rarely) galloping. The vast amount of footage necessitated many months of work in analysis, but this should be done for every species in any case, because even within one gait the movements of the legs may be significantly different at slow and fast speeds.

6.2. *Gaits and Gait Analysis*

The study of gaits of animals moving on four legs is necessarily more complicated than that of man, because now we are considering twice as many legs, and legs which often move in more than one plane. Fortunately, depending on the gait being used, the four legs work together in more or less regular patterns. These include the slow *walk*, the medium speed *trot* and *pace*, and the fast *gallop*. Almost without exception, the gaits of quadrupeds are studied using moving picture films, because the eye alone is not able to distinguish and describe the various movements of the legs. Slow motion film is necessary if the animals studied are small or move fast. The basic information usually gleaned from each frame of a sequence of animal movement is which legs are supporting the animal at any one instant. The ones that are not supporting the animal are being swung forward preparatory to the stance position. For convenience these data can be written in the form shown in fig. 6.1. These various support combinations of legs follow each other in definite patterns depending on the gait. Thus a typical slow walk of a hoofed animal (described first because such animals are common and large, with long legs that move relatively slowly) can

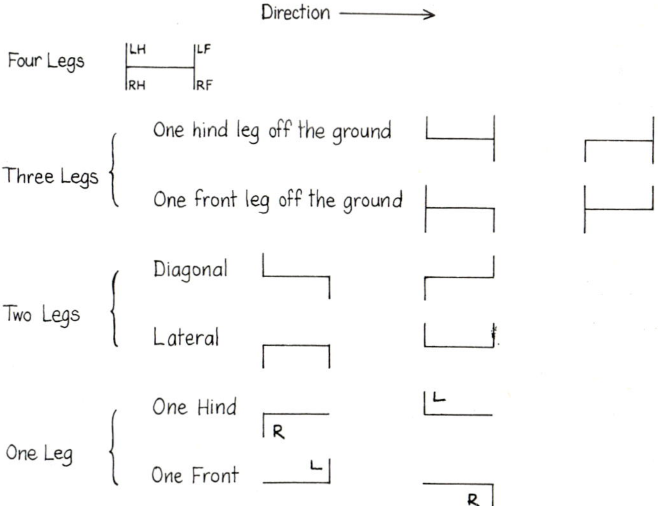

Fig. 6.1. Combinations of supporting legs of quadrupeds. The horizontal lines represent the body of an animal, the vertical lines its legs. L—left; R—right; H—hind; F—front.

be depicted by supporting legs as in fig. 6.2(*a*). If the walk becomes very slow, four legs may support the animal at intervals during each stride, often replacing the stance of two diagonal supporting legs (fig. 6.2(*b*)). The order in which the legs swing forward is always the same: left hind, left front, right hind, right front, i.e. *lateral sequence*, where a front leg is set down after the hind leg on the same side. The animal is never supported by a single leg, nor by two front or two hind legs alone as it is in faster gaits.

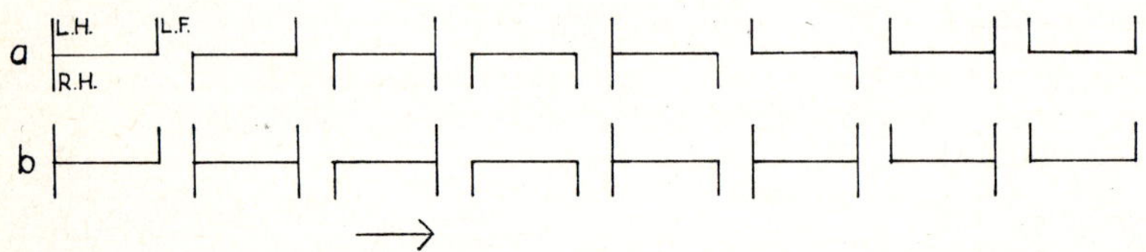

Fig. 6.2. Combinations of supporting legs in sequences of a walking hoofed animal. (*a*) Slow walk; (*b*) Very slow walk. L—left; R—right; H—hind; F—front.

So far a gait's description has entailed noting in what order the four legs are set down and lifted. The description can be refined by detailing how long each leg or combination of legs supports the animal during a stride. Differences in the walk of various species can be found by calculating the percentage of time during a stride during which each combination of legs is on the ground. When this is done it is evident that ruminants have a *walk pattern* which is similar for the animals within each family (table 6.1). The camelids and giraffids use lateral supporting and four supporting legs in their walk more than do bovids and far more than do cervids, whereas the cervids make great use of diagonal supporting legs. The most diagnostic characteristic of the walk pattern is the use of diagonal as opposed to lateral legs when only two legs are supporting an animal. When lateral legs are used to a large extent, the two legs on one side of the animal swing forward more or less together; when diagonal supporting legs are predominant, the hind hoof on one side is placed down so close to the forehoof on that side that there appears to be danger of a collision (fig. 6.3). Lateral legs support most species for at least some part of the stride, since the hind leg on one side of the body is placed on the ground about where the front leg was located, and during this exchange the animal must be supported only by the legs on the opposite side. Diagonal legs may virtually never be used, as in walking giraffe and okapi, animals which have long legs but relatively short bodies and whose legs might collide during a diagonal stance pattern of leg use.

TABLE 6.1. *Average percentage of time during a stride spent by ruminants, walking on flat ground, on supporting combinations of legs. After A. I. Dagg and A. de Vos (1968) Journal of Zoology (London), **155**, 103–110; (1974) Journal of Zoology (London), **174**, 67–78.*

Family and species	Lateral legs on ground	Diagonal legs on ground	One hind leg off ground	One front leg off ground	Four legs down	Number of strides averaged
Camelidae						
Camel *Camelus dromedarius*						
under 6 months	44	0	22	25	9	40
yearling	41	0	26	28	5	58
adult	33	0	30	34	3	387
adult pulling water	22	7	34	37	0	56
Cervidae						
Fallow deer *Dama dama*	1	25	30	41	3	183
Sika deer *Cervus nippon*	3	21	32	43	1	18
White-tailed deer *Odocoileus virginianus*	5	18	38	39	0	212
Mule deer *Odocoileus hemionus*	6	20	35	39	0	32
Caribou *Rangifer tarandus*	7	22	33	38	0	113
Père David's deer *Elaphurus davidianus*	8	2	44	42	4	8
Red deer *Cervus elaphus*	10	20	34	36	0	21
Wapiti *Cervus canadensis*	10	21	34	35	0	179
Moose *Alces alces*	10	12	41	36	1	20
Giraffidae						
Giraffe *Giraffa camelopardalis*	40	0	20	25	15	92
Okapi *Okapi johnstoni*	28	0	33	32	7	18
Antilocapridae						
Pronghorn *Antilocapra americana*	20	7	30	41	2	20
Bovidae						
Saiga antelope *Saiga tatarica*	30	4	25	37	4	18
Grant's gazelle *Gazella granti*	32	4	29	35	0	9
Lechwe *Kobus leche*	33	3	28	33	3	34
Tsessebe *Damaliscus lunatus*	32	8	27	33	0	9
Wildebeest *Gorgon taurinus*	26	4	32	35	3	13

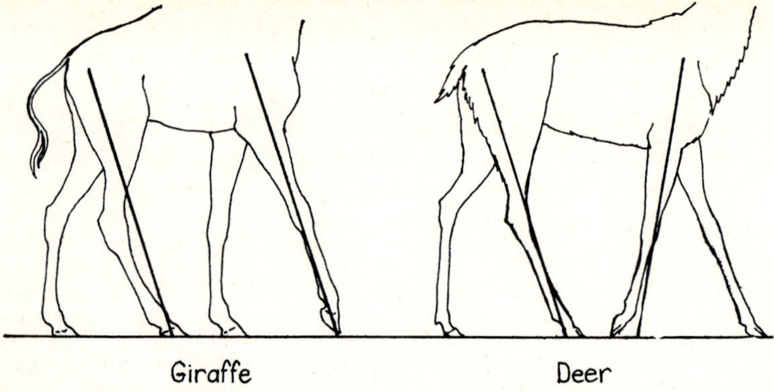

Fig. 6.3. Walking giraffe and deer. When the right hind foot has just been placed on the ground, the right front leg is completing its swing in the giraffe, but just starting its swing in the deer.

A method of quantifying all symmetrical slow or medium-speed gaits, the trot and pace as well as the walk, is the *gait formula* of Dr. M. Hildebrand, which takes into account two measurements: the percentage of the stride-interval that each hind foot is on the ground, and the percentage of the stride-interval between the footfall of the forefoot

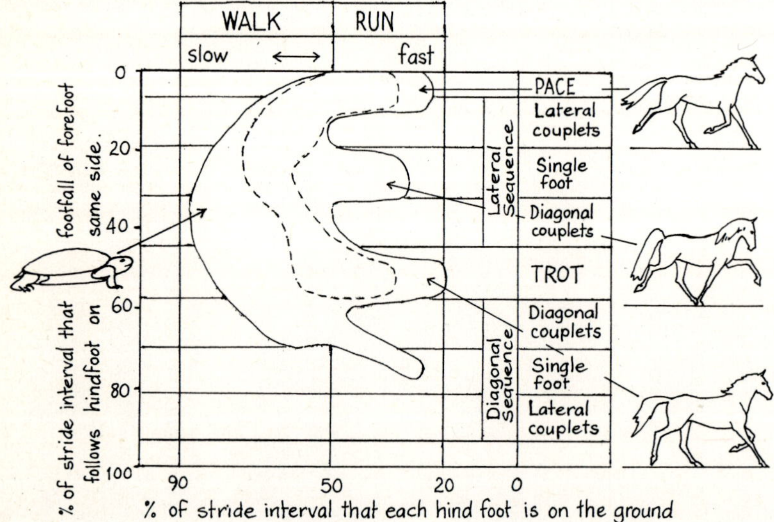

Fig. 6.4. Plot of the limits of the gait formulae of 158 genera of quadrupeds (inside solid line) and 37 breeds of dogs (inside dashed line). The names of various symmetrical gaits are given in the right hand margin. Based on the work of M. Hildebrand (1968), *Journal of Morphology*, **124**, 353–360.

and that of the hind on the same side. This formula varies with the speed of an animal, because a foot generally supports an animal for a shorter period the faster the animal moves, but it is often characteristic of a species or breed. In fig. 6.4 Hildebrand has outlined the figures inside which the gait formulae of tetrapods of 158 genera (solid line) and 37 breeds of dogs (dashed line) fall. From the larger figure it is evident that a gait can be so slow that each hind foot is on the ground up to 88 per cent of the stride interval, or so fast that each is on the ground less than 20 per cent of the stride interval. The front leg and hind leg on one side of an animal can swing forward together (pace at top of figure), or the footfall of the forefoot can follow nearly 80 per cent after that of the hind foot. The timing of the use of the front and hind legs has been given a variety of names (to the right of figures), but these are not in common use.

The gaits of dogs indicated in fig. 6.4 are elaborated in fig. 6.5. The German shepherd or Alsatian and bloodhound occupy the upper right corner because they are pacing, the boxer and Great Dane the lower right corner because they are trotting. The second Great Dane is near the top of the figure because it is walking (i.e. each hind foot is on the ground more than 50 per cent of each stride). Those dogs to the left of the figure are moving slowly. No dog swings a right front leg forward immediately after a left hind leg in a symmetrical gait, and none uses a single foot gait (see fig. 6.4) when going fast.

6.3 *Walk*

The walk is a slow, regular, symmetrical gait in which the left legs perform the same movements as the right legs but half a stride later and in which either two, three, or four legs are supporting the animal at any one time (as fig. 6.2; fig. 6.6). The similar walk patterns within each artiodactyl family (table 6.1) are probably mostly based on anatomical similarities—of the joints, muscles, bones and nervous system. For example the walk of the giraffids may be correlated with their relatively long legs in proportion to their body length. An extended use of lateral supporting legs is essential if the front leg on one side of the body is not to curtail the forward swing of the hind leg on the same side. In addition, unlike most quadrupeds the two giraffids have longer forelegs than hind legs. However, the distinctively long use of lateral supporting legs cannot be correlated solely with these factors, because extensive use of lateral legs has also been found in adult camels (33 per cent of time of stride) and in camels under six months of age (44 per cent of time of stride). The relatively long use of lateral legs as supports which has evolved in some higher quadrupeds may allow several advantages (although it reduces stability):

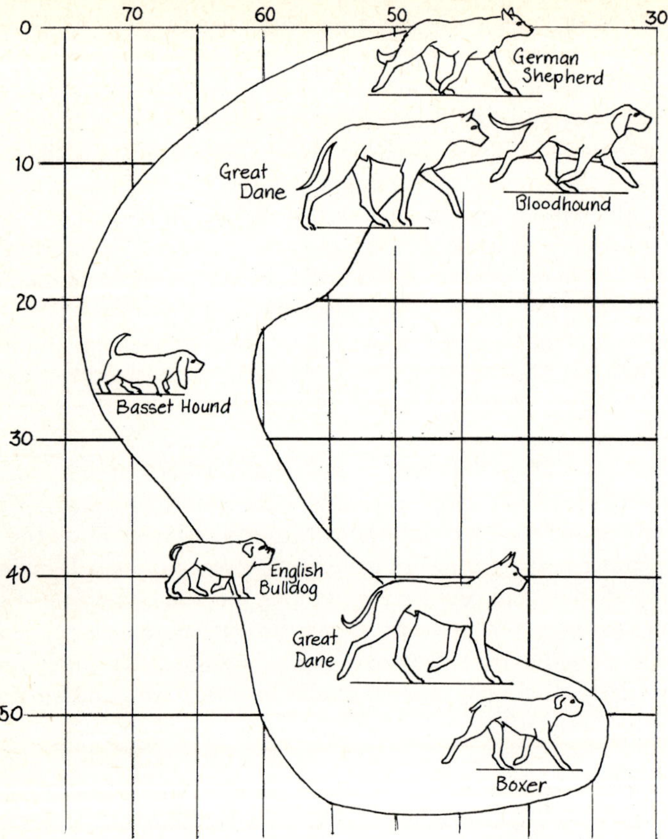

Fig. 6.5. Plot of the gait formulae of dogs (as in fig. 6.4), with the position of various dogs indicated. Based on the work of M. Hildebrand (1968), *Journal of Morphology*, **124**, 353–360.

(*a*) lateral legs swinging forward more nearly together may enable a greater use of the trunk musculature and an expenditure of less energy;

(*b*) there is no danger of the hind leg coming in contact with the front leg on the same side as it swings forward;

(*c*) perhaps because of the foregoing possibilities, the walk using lateral legs allows longer strides and therefore a faster walk. (Although not homologous it might be noted here that pacing horses are slightly faster than trotting ones).

Any quadruped is more stably balanced if it is supported by three legs rather than by two, and by diagonal legs rather than lateral legs if only two are in use. From table 6.1 it is apparent that the walk patterns

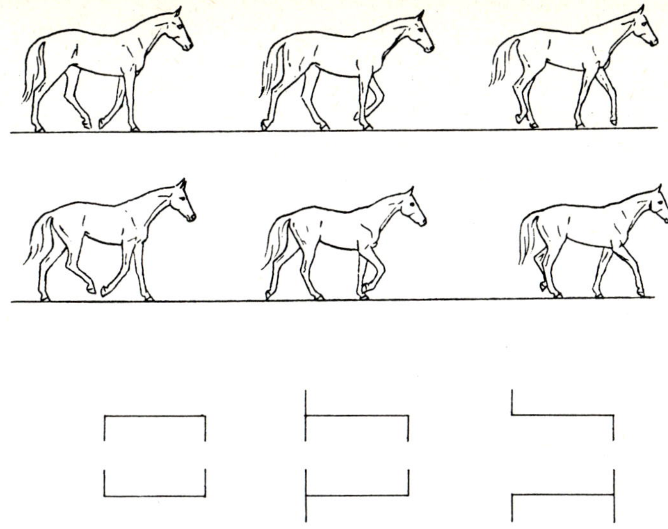

Fig. 6.6. Sequence of a horse walking, with support diagrams.

of the Cervidae, a woodland group, are much more stable than those of the pronghorn antelope or the true antelopes, the representatives of which, although roughly similar to cervids in size and shape, all live in open areas. It is possible that the extended use of diagonal legs in the cervids is directly related to their quite dense habitats. By having a more stable walk cervids are able to pause more quickly and to leap away more readily if danger threatens. Such danger is more immediate in an environment where vision is limited than in open country. A stable walk is also the primitive type; quadrupedal amphibians and reptiles use diagonal legs to support themselves rather than lateral legs if two legs are used at once.

If we exclude the red deer which has only lived in open areas in Britain during historic times, the only other cervid which spends much of the year in open habitats is the barren-ground caribou or reindeer, *Rangifer tarandus*, from the vast expanses of northern Canada and of northern Eurasia. The caribou is difficult to study because it rarely lives long in captivity and when it is filmed in the wild in winter, the low temperatures may make the camera speed slow down during the filming. It might be expected that because of its tundra habitat, this animal would use a less stable walk than other cervids. The fact that it does not may be related to its relatively recent immigration during the Pleistocene onto the tundra from the temperate forests. As well, its walk pattern is probably mostly adapted to winter conditions when the caribou moves into the forests bounding the tundra and when the concept of survival of the fittest is most poignant.

Of the giraffids, the okapi has the more stable walk as well as the denser habitat. However, the walk patterns of both the giraffe and the okapi, with the long use of lateral supporting legs, are apparently primarily dependent upon their anatomy and only secondarily upon their habitat.

The filmed walking gaits of individuals will vary slightly, depending on irregularities in the terrain, the animal's mood, an individual's stiffness, a sound which may make it pause for a fraction of an instant, or the bite of an insect which may disrupt the animal's movements. As well, variations in reading the supporting legs in each photographic frame are inevitable since, although the animal may be walking smoothly, its movements are only recorded at fixed intervals. These variations will average out over a large number of frames. However, there are other variations that are not random, one of which is the speed of the walk. The walk pattern at a fast speed may differ significantly ($P < 0.01$) from that of the same species at a slow speed. At faster speeds the cervids use diagonal legs more and three supporting legs less; the pronghorns use diagonal legs more and lateral legs less; and the giraffids use lateral legs more and three or four legs less.

Stable walking patterns (those using diagonal rather than lateral pairs of legs, and/or three legs rather than two) tend to be predominant in young animals and in those which are moving through rough terrain—such as caribou manoeuvring among logs and rocks in the wild, or moose traversing shallow water or snow. They are also predominant in individuals with heavy horns or antlers, compared with members of the same species without such headgear, for example antlered bull moose, male lechwe, and antlered caribou of either sex; and in camels pulling up heavy loads of water at desert wells.

Most small mammals and other tetrapods use the lateral sequence walk. Some bats use this method, walking with their diagonal legs moving almost together, their weight being borne on their hind feet and thumbs, and sometimes on their tails. Sometimes the thumbs are set down in the same line as the hind legs, and sometimes outside them. The vampire bat (*Desmodus rotundus*) is perhaps the most agile bat of all; it can raise its body high off the ground and maintain an almost upright posture while moving about unobtrusively on the bodies of the animals on which it feeds (fig. 6.7).

The primates are one of the few groups of mammals which do not use the sequence of legs noted in fig. 6.2, perhaps because of their unusual morphology, which generally includes well developed shoulders plus long front and short hind legs. They generally move the opposite

Fig. 6.7. Vampire bat in walking stance. These bats can run nimbly over the bodies of their sleeping victims. Based on the work of M. J. Lawrence (1969), *Journal of Zoology* (London), **157**, 309–317.

foreleg after a hind leg in *diagonal sequence*: left hind, right front, right hind, left front—the *pithecoid* walk of Muybridge. Primates are nearly unique in that each hind foot is passed to the same side of its respective forefoot, which orients the spine at an angle to the line of progression, as crabbing does in dogs. Thus often one shoulder comes to be ahead of the other during the walk. Non-primate mammals that use the same diagonal sequence are the aardvark (*Orycteropus afer*), the kinkajou (*Potos flavus*), and the giant armadillo (*Priodontes giganteus*), but as they have relatively short legs which would not strike each other at moderate speeds these species do not exhibit asymmetrical crabbing.

The quadrupedal walk is common to all mammals with the exception of some that have hind legs much more developed than forelegs and some that are entirely arboreal. Except for a few primates, mammals with unusually large hind legs nearly always move these legs simultaneously and not alternately, either by themselves or in conjunction with the front legs. Such mammals include kangaroos, wallabies, small jumping rodents, spring hares (*Pedetes*) and chinchillids. Although ancestral animals in the basic kangaroo stem evolved as hopping terrestrial forms, the tree kangaroos (*Dendrolagus*), unlike other macropodids, sometimes use a quadrupedal walk on branches or along ledges in zoos. Similarly, whereas most lagomorphs hop or bound, some occasionally walk. This is a not uncommon gait in rabbits that must manoeuvre on mud, such as the swamp rabbit *Sylvilagus aquaticus* and the marsh rabbit *S. palustris*; a soft substratum is a poor one for a hopping gait which necessitates a large downward thrust on a small area. Lagomorphs that can walk but rarely do so include the black-tailed jackrabbit (*Lepus californicus*), the white-tailed jackrabbit (*Lepus townsendii*), the snowshoe rabbit (*Lepus americanus*), the cottontail

(*Sylvilagus floridanus*) and the European rabbit (*Oryctolagus cuniculus*). The arboreal group of mammals that do not use a quadrupedal walk includes the sloths and gibbons. Since these species seldom move any distance on the ground, they possess no other terrestrial gaits either.

A *running walk* or *amble* is one in which the footfalls are the same as those in the walk but in which the speed is faster, and it cannot be sustained for long periods. This gait may be common in small species, but it can be identified accurately in them only in slow-motion films, so that little is known about it. It is an artificial or taught gait in horses, and the normal fast gait of elephants. Because of their mass, elephants are usually supported by at least one front and one hind leg. They are unable to jump so they have been safely confined behind a ditch only 1·7 m wide and 1·4 m deep.

A final walking gait is that of species such as the kangaroo rat (*Dipodomys*), kangaroos and wallabies, and possibly pangolins (*Manis*), which have well developed tails that the animals use for support as a fifth appendage during slow locomotion. These animals swing their hind legs forward while their weight is supported on their front legs and tail.

Although virtually nothing has been done on quadrupeds to compare with the in-depth work done on man, W. C. Hutton, M. A. R. Freeman, and S. A. Swanson of London have studied the dog's walk using a strain-gauge force-plate. They found that the maximum vertical force recorded for the front pads was approximately equal to the dog's body weight, while for the back pads this force was only about 0·8 times this. This is as expected, because the centre of gravity of a dog lies nearer its front legs than its hind legs. The forces of the walk in the horizontal direction were about one tenth of the corresponding vertical forces. The times of contact of the front pads were as much as 1·5 times those of the back pads, and shorter in the short-legged beagle than in the long-legged labrador. When the speed of the walk increased noticeably, the maximum forces exerted by the dog's legs increased, while the contact time of the feet on the ground decreased.

Many decades ago it was calculated that the walking human leg did not swing passively forward as would a pendulum, and the same is true for the legs of quadrupeds. If l is the length of a leg, and its period of swing is considered to be the same as that of a uniform cylindrical pendulum—which is $\pi\sqrt{(2l/3g)} \simeq 0\cdot8\sqrt{l}$ second if l is in metres—this works out to be roughly twice the actual time of swing of the walking leg as calculated from moving picture sequences. Thus the leg muscles play an important role in pulling each leg forward, even in a slow walk. As would be expected, animals with longer legs have a slower natural frequency of swing of their legs than do species with shorter legs.

6.4. *Trot*

The trot is a symmetrical gait of intermediate speed in which two diagonal legs usually support the body when it is in contact with the ground (fig. 6.8). The two diagonal legs are not necessarily set down upon the ground at the same instant. Usually, if the beats are not quite synchronous, the feet touch the ground in the same order as in the walk: left front, right hind, right front, left hind. In the "collected trot" of a trained horse however, in which the legs are raised high and placed well under the body and in which there is much action but little forward progression, the sequence is right hind, left front, left hind, right front.

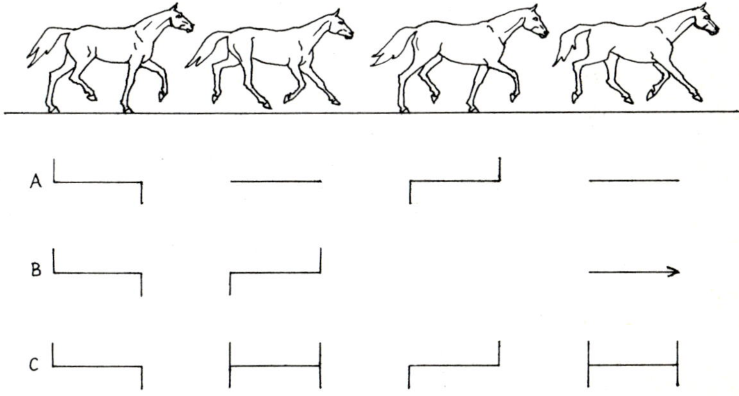

Fig. 6.8. Sequence of a horse trotting quickly, with support diagrams for this sequence (A) and for slow trotting (B, C).

The trot is an enduring gait restricted to animals with front and hind legs of similar length. It is not present in any of the species with greatly enlarged hind legs, nor is it found in some species with sloped backs such as the ring-tailed lemur (*Lemur catta*), spotted hyena (*Crocuta crocuta*), giraffe, possibly okapi (the London Zoo and the Brookfield Zoo in Chicago have exhibited okapis, but their keepers are not sure whether they can trot or not), and maybe bears. (It is amazing that no-one is sure exactly how bears run). The trot is found in other slope-backed animals however such as the wildebeest (*Gorgon taurinus*) and hartebeest (*Alcelaphus*), and in viverrids in which the front legs are shorter than the back. The back must be fairly straight, as the propulsive force comes from the right hind quarters and left forequarters (or *vice versa*) at one time. Animals such as bears, agoutis (*Dasyprocta*), apes and monkeys which do not trot may have a spine that is not straight and rigid enough; in apes the back and pelvis often rotate sharply from side to side during the symmetrical walk, so as to increase the length of stride of the relatively short hind legs. Nor can the body be very long, as it is in most sciurids and small mustelids; mustelids that are not too long to trot are the wolverine (*Gulo gulo*), European badger (*Meles meles*), ratel (*Mellivora capensis*), and tayra (*Tayra barbara*). Very small animals such as voles (e.g. *Microtus pennsylvanicus*) can apparently trot, judging from my films, although some workers believe that small mammals only use a fast walk and a bound or gallop. Further research needs to be done on gaits in small mammals. Large animals such as the rhinoceros and the hippopotamus can trot, whereas the huge elephants cannot.

6.5. *Pace*

This symmetrical gait is one of intermediate speed in which two lateral legs usually support the body when it is in contact with the ground (fig. 6.9). Pace, not rack, is the accepted word for this gait in harness racing of horses around the world, so it is best to use this word for the same gait in other quadrupeds too. As in the trot, animals that pace must have front and hind legs of roughly the same size. Animals with sloping backs such as giraffe and hyenas that move their lateral

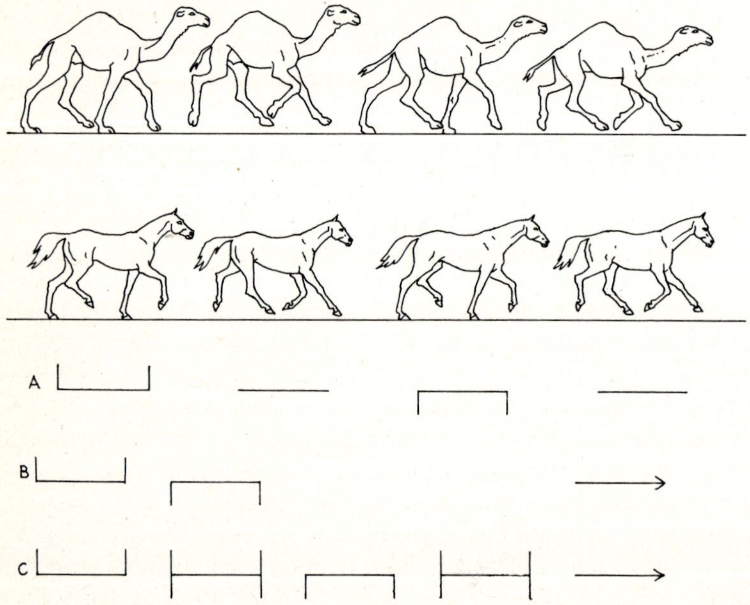

Fig. 6.9. Sequence of a camel and a horse pacing quickly, with support diagrams for the horse (A) and for slow pacing (B, C).

legs almost together in the walk, still do not pace. The animals that do pace are the dog, horse, camelids and perhaps bears. Since this rare gait is present in the two domestic animals that have been bred most intensively by man for centuries, it may be that the pacing gait is heritable in many species but that it is useful only in mammals with suitable anatomy, or in those such as camels which inhabit flat terrain, since the pace is a less stable gait than the trot. In horses the use of the pace is governed by a single gene which is recessive to the dominant gene which governs the trotting gait. Thus in horses natural pacers are homozygous recessives that breed true. However, trotting horses can be taught to pace with the use of special shoes and hobbles. Among dogs, only long-legged breeds pace. These include the bloodhound, German shepherd or Alsatian, golden retriever, collie, great Dane, Rhodesian ridgeback, saluki and weimaraner.

Because of its instability, the pace is only useful on flat ground; it has undoubtedly developed in the camel because of the flat desert habitat of this species. It has also been filmed briefly in the llama (*Lama glama*), but the pace is not present in the vicuña (*Vicugna vicugna*) which is smaller and may occupy more rocky habitats in the Andes Mountains. A suitable anatomy for the pace is:

(*a*) a fairly large size, so the mammal will not roll too far sideways while supported by lateral legs (large dogs are minimum in size);

(*b*) a slim build, so that the centre of gravity of the animal can more or less be shifted over these supporting legs (camels, horses and dogs); or

(*c*) heavy limbs to provide a sturdy support and a low centre of gravity (bears?).

The pace and the trot are almost equally fast in the horse, but the pace is a superior gait in that the front and hind legs never hit each other as may happen in the trot.

6.6. *Gallop and Bound*

The *gallop*, sometimes also called the canter or lope, is the fastest gait of most animals, with the body often unsupported following a push-off with the front legs (*flexed suspension*) and sometimes with the hind legs (*extended suspension*). It is usually asymmetrical, with right and left legs doing different movements in a stride. If the last hind leg to touch the ground is followed by the front leg on the same side, it is a *rotatory*, *rotary* or *lateral gallop* (fig. 6.10); if it is followed by the front leg on the opposite side, it is a *transverse* or *diagonal gallop* (fig. 6.11). Not all galloping strides are asymmetrical. Gambaryan reported finding the footprints of a young Pamir argali (*Ovis ammon*) which was going so fast that it was in a state of suspension after the contact of each foot

Fig. 6.10. Rotary gallop of a greyhound with support diagrams. The dog has an extended period of suspension (no. 3) and a flexed period of suspension (no. 6).

Fig. 6.11. Transverse gallop of a horse with support diagrams. The horse has only a flexed suspension (no. 6) in its gallop.

on the ground, giving the fastest possible support formula of 1–0–1–0–1–0–1–0.

The gallop may intergrade with the *bound* in many species, or with the *half-bound* (fig. 6.12). In the bound the push-off is always with both hind feet and the landing on both front feet. In the half-bound an animal pushes off as in the bound but lands with one front foot before the other, a sequence which decreases the forward velocity of an animal less than the bound does.

Because the gallop in general has an asymmetrical use of the four legs, there is no way in which a compact formula such as the *walk pattern* or

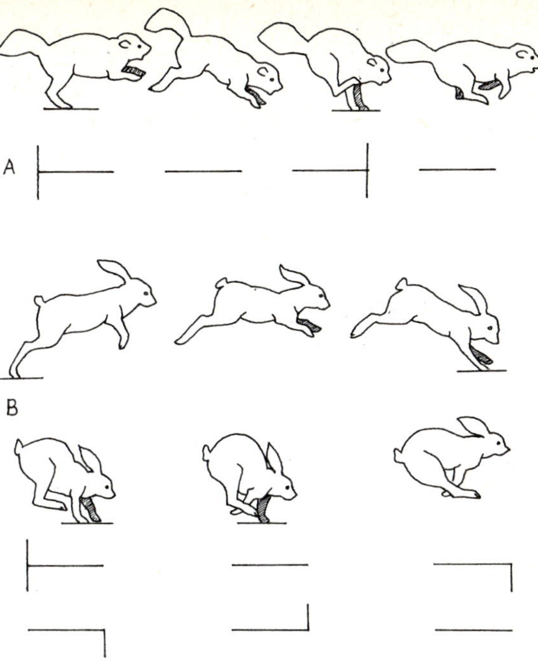

Fig. 6.12. Bound of the Siberian souslik (*Citellus undulatus*) (A) and half-bound of the European hare (*Lepus europaeus*) (B) with support diagrams. After P. P. Gambaryan (1972), *How Mammals Run*. U.S.S.R.

gait formula can be applied to describe the leg movements. Instead, a chart can be made of successive strides which depicts how long each leg supports an animal and how the legs move in relation to each other (fig. 6.13). The erratic gallop of the wildebeest has often been described, and when I analysed in detail the gait of one galloping in a straight line, I realized that the irregularity of the sequence was caused by frequent changes of leads of the feet. A change of lead occurs when, in successive strides, the same front or same hind leg is not set down before its pair upon the ground. In the first two strides of the wildebeest, the lead foot was the right front, which was the last to leave the ground before the period of suspension (fig. 6.13). In the next six strides the lead changed to the left front, to the right front, and back to the left front foot again. Whereas most quadrupeds change leads only after many strides, when the lead foot becomes tired from the extra stress of pushing the animal into its period of suspension, the wildebeest does so for no apparent reason. The hind legs too changed their order of being set down twice in the eight strides. As horse-riders know, a change of lead occurs when an animal is turning a corner as well as when its lead leg is tired. If the

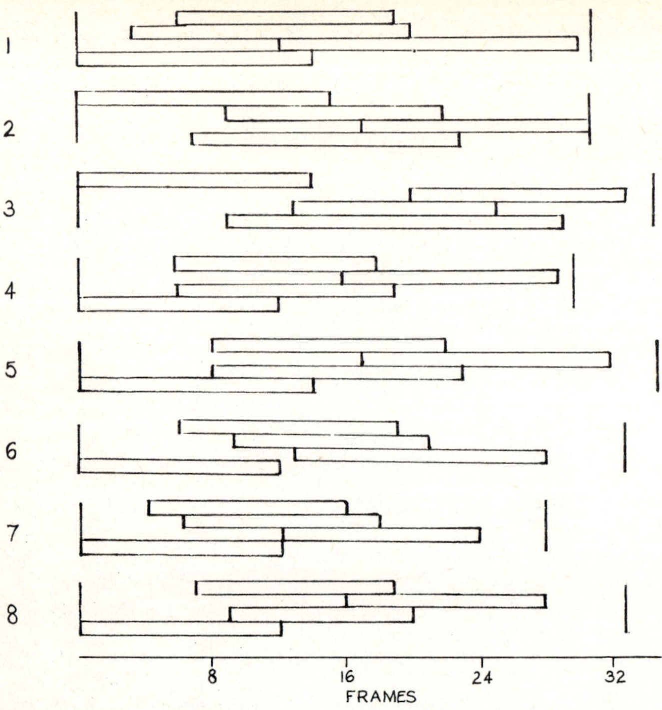

Fig. 6.13. Eight successive strides in the gallop of the wildebeest. Bar diagrams are given of the length of time each foot is on the ground, in the order left hind; left front; right front; right hind. After A. I. Dagg (1969), *Journal of Mammalogy*, **50**, 825–826.

animal is turning to the right, it will lean into the turn and lead with its right front leg, as this leg will be most nearly under its centre of gravity.

The gallop and bound are almost universal in terrestrial mammals because they are used to catch prey or to escape from predators. They are considered together because they are closely allied, although the bound is a symmetrical and the gallop usually an asymmetrical gait. For example, cheetahs (*Acinonyx jubatus*) tend to gallop when travelling slowly and to bound when chasing prey; galloping antelope may give occasional high bounds as they flee, to see better or to confuse their predators; and the gait of a marsupial mouse has been described as a "bounding gallop". Species which only bound and do not gallop have flexible backbones, such as small sciurids, jumping mice (*Zapodidae*), long mustelids, and the noctule (*Nyctalus noctula*). Tree-climbing species in general tend to bound, perhaps because arboreal mammals are usually light. Large mammals sometimes bound too if they are bogged down in heavy snow or mud. Species which only gallop and do not bound tend to have short bodies and quite rigid backbones, such as ungulates, canids, ursids, hyenas, large sciurids such as the woodchuck (*Marmota monax*), and large mustelids such as the striped skunk (*Mephitis mephitis*) and the wolverine. Perhaps because of an inflexible backbone and a low-slung body some primitive quadrupeds such as the opossum (*Didelphis marsupialis*) and the platypus (*Ornithorhynchus anatinus*) are unable to gallop or bound. As with the trot and pace, mammals that gallop must have front and hind legs of similar size, although the differences in length may be greater. In the macropodids most kangaroos and wallabies do not bound but the tree kangaroos (*Dendrolagus*) and quokkas (*Setonix brachyurus*) do. These species have relatively longer forelegs than do the other kangaroos.

Some mammals are so heavy that they cannot hoist themselves into the air for a gallop. Thus, elephants (max. 7000 kg) and hippos (max. 3000 kg) do not normally gallop, but rhinos (max. 2000 kg) do. (Similarly Malaysian crocodiles less than 2 metres long can gallop, whereas larger ones never do.) Even when hippos are moving underwater where they are buoyed up by the water, they trot rather than gallop, implying that a gait which has never been learned would not suddenly be used.

6.7. *Choice of Fast Gait*

Many mammals can both gallop and trot, but they habitually use one gait rather than the other to cover distance quickly. Large animals such as the moose, wapiti, caribou, eland (*Taurotragus oryx*) and waterbuck (*Kobus*) tend to trot rather than gallop at speed. The trot may be preferred for any of the following reasons:

(*a*) It is less tiring, although the gallop is faster.

(*b*) The trot is more stable, since two diagonal legs support the animal for most of each stride. One leg rarely supports an animal by itself as it often does in the gallop. Thus the trot is a better gait on mud or snow or along a rough trail in thick vegetation.

(*c*) The centre of gravity changes less drastically during the trot than it does during the gallop. Periods of suspension in a gallop are in fact leaps, which require considerable energy to raise the centre of gravity of an animal. Long leaps are necessarily high ones and require more energy than short ones. Large animals, especially those with heavy horns or antlers, would find a gallop particularly tiring, since the headgear would have to be lifted higher during each stride in the gallop than in the trot.

6.8. *Energy Relationships*

Unlike reptiles which increase their speed significantly by increasing the length and frequency of their strides within their single gait, cursorial mammals do so by changing their gaits from low gear (walk) to second gear (trot or pace) to high gear (gallop). The speed of these gaits is not absolute, but depends on an animal's size; at 6 km h^{-1}, for example, a horse is walking slowly, a dog is trotting, and a mouse is galloping. The point of change from one gait to the next appears to be a function of an animal's mass. Dr. C. R. Taylor and his colleagues at Harvard University have postulated that the transition from the trot to the gallop occurs at "physiologically similar speeds" for animals of different size. At each gait some elastic energy is stored during each stride—little during walking; more during trotting, largely in the elastic elements of the legs; and most during galloping, both in the legs and in the entire trunk. For example, when a horse's foot hits the ground, the impact bends the fetlock joint between the metapodial bones and the phalanges, stretching the suspensory ligaments there. These ligaments are elastic, so the energy of deformation is recovered as the horse's foot is lifted for the next stride. The faster a horse goes, the more vigorously it places its feet on the ground, the greater is the energy of deformation, and the greater the spring the animal has after each step. It may be that additional energy is stored in the body at the transition of one gait to another.

The contraction of muscles during galloping produces heat which tends to accumulate in an animal's body, sometimes to such an extent that the animals are forced to rest. To study this phenomenon, C. Taylor and V. J. Rowntree of Harvard University made two cheetahs and two goats run on treadmills in rooms held at various temperatures. Each animal wore a ventilated mask which supplied air and by which oxygen consumption and hence heat production were calculated. Heat storage within the animal was obtained by noting changes in its rectal temperature, while the evaporative heat loss was calculated from the weight loss of the animal (the figures excluded the animals which urinated or defecated during the running). During sprints, heat production in an animal may increase by more than 50 times, and be nearly five times the maximum rate of heat absorption from the environment in a hot desert. Taylor and Rowntree found that the heat production of running cheetahs, as of other animals, increased linearly with speed. A cheetah running for 15 minutes at 11 km h^{-1} stored 70 per cent of the heat it produced, and at 18 km h^{-1}, 90 per cent. The values for heat storage were much lower for the goats, which are non-sprinters, and indicate that sprinters do not necessarily develop evaporative cooling mechanisms for maintaining a constant body temperature during running. The cheetahs refused to run when their body temperature reached 40·5°C. Instead they lay down with their feet in the air and slid on the tread surface. Perhaps in the wild the length of time a cheetah can chase its prey is determined by the amount of heat its body can store before reaching a limiting body temperature. Gazelles, like cheetahs, can store large amounts of heat when they run at high speed, but unlike the cheetah the gazelle's body temperature may increase by 5–6°C during a run. (The temperature of the brain does not reach such high levels because it is selectively cooled via a countercurrent heat exchanger in the carotid rete.) Thus if a gazelle can avoid capture during a cheetah's initial burst of speed, perhaps there is a good chance that the cheetah will overheat and stop running before its prey does.

The energetic cost of locomotion for four-footed mammals is a function of body mass, with large mammals needing less energy per unit of body mass than do small mammals, as one might expect from the lower rate of metabolism of large animals. Bipedal animals do not necessarily fit into this relationship; for rheas (*Rhea americana*) and men the energy cost of running is about twice what it would be for a quadruped of the same size, although the reason for this is not clear. Is the extra cost in energy of moving on two legs rather than four the price man has had to pay for freeing one set of limbs for other activities? Only when bipedal and quadrupedal animals each weigh about one kg is the predicted energy cost for each of running one kilometre equal.

6.9. *Occasional Bipedalism*

The alternate bipedal gait used by man has been discussed at length in chapter 4; a similar gait, either as a walk or a run, is used occasionally by long-legged prosimians, monkeys, apes, and the African pangolin (*Manis*). The capuchin (*Cebus capucinus*) and spider monkey (*Ateles*) are particularly active bipedally, the latter running with its trunk leaning forward and its tail held out backwards for balance, or with its arms raised above its head and its tail also raised vertically, since a vertical line from the centre of gravity then runs down through the body. A few mammals with well developed tails may also use an alternate walk with their trunks held vertically; these animals, which include pangolins, kangaroo rats, and jerboas, use it when moving slowly, dragging their tails along the ground behind them.

6.10. *Display Gaits*

Display gaits serve not only to transport an animal from one place to another, but also to communicate its emotions to others. They do not include non-continuous movements like the stamping of feet to warn of danger or the excitation jump of the caribou. The best known is the *stott*, also called the pronk or spronk, which is performed with all four legs taking off and landing together. During the period of suspension the legs hang down vertically from the body. The stott is only found in gregarious pecoran species, but it is widespread in them, occurring in at least sixteen species including the mule deer, pronghorn, gazelles and lechwe. Although the stott is slower than a gallop, it is effective in changing direction and in climbing hills. It presumably acts in intra-specific communication by allowing the pronking animal to be seen by others, particularly if it has a distinctive rump or tail as it usually does; by making a distinctive noise which may warn of danger; and/or by often depositing scent on the ground from glands in the feet. It also gives the fleeing animal a good view from the height of the bound. Since the hooves are set down close together, this gait may help prevent them from becoming entangled in dense vegetation and offer the animal a rapid start.

Other display gaits occur in gregarious animals such as some primates and lions, in which dominant males for example may strut or swagger. In gregarious ungulates a territorial male may *prance*, moving stiff-legged with short, accentuated steps, to communicate its status to others in the area, while other mammals will walk with stiff, high steps preparatory to charging an enemy. Even human beings are not exempt; Isaiah (3: 16) mentioned with disapprobation certain haughty daughters of Zion who "walk with stretched forth necks and wanton eyes, walking and mincing as they go, and making a tinkling with their feet". Too few display gaits have been described in detail, but when this has been done a comparison of the movements should be of great interest from an evolutionary point of view.

6.11. *Neck Movements*

So far we have discussed in the locomotion of quadrupeds only the movement of the legs. The neck also has a role to play, and in no animal is this easier to study than, needless to say, in the giraffe. From films I had taken in South Africa I traced the outlines of giraffe as they walked and galloped. These tracings showed that there was a close correlation between the position of a giraffe's neck and the phase of its stride. When a walking giraffe had all four feet on the ground, the neck was moving backwards (fig. 6.14, A and F). As two legs on one side of the body left the ground (B and C), the neck reached its most posterior point (= lowest point on the neck-back angle curve plotted in fig. 6.14) and began to swing forward again, thus shifting the centre of gravity of the giraffe forward along with the legs. Just before the two right or two left legs were set on the ground again (E), the neck reached its farthest point forward (= maximum neck-back angle) and began to swing back, thus helping to decelerate momentarily the giraffe's movement forward. Thus the neck moves back and forward like a pendulum during walking, making two cycles for each stride. By pressing forward at the beginning

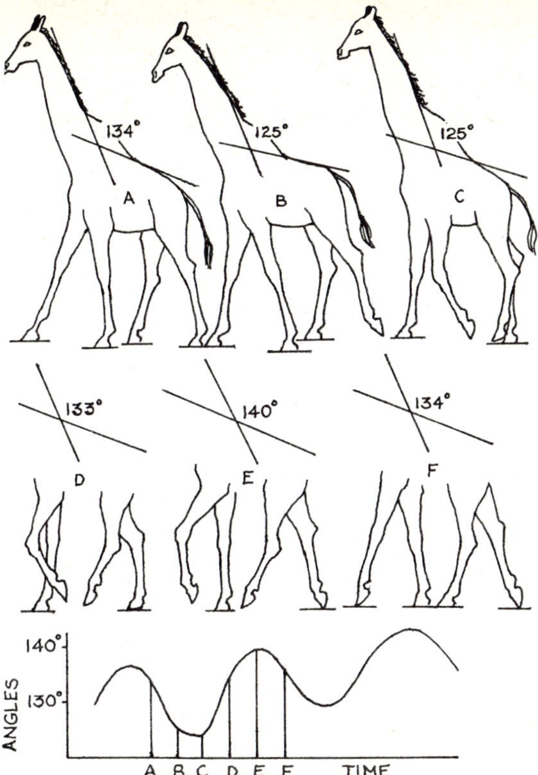

Fig. 6.14. Half a walking stride of a giraffe. The angles that the slope of the neck make with the back are given beside each giraffe and are also plotted against time. After A. I. Dagg (1962), *Journal of Mammalogy*, **43**, 88–97.

of each half-stride, the neck moves more into line with the power stroke. At the end of each half-stride, as the hooves touch the ground again, the neck swings backward in order to slow down the forward momentum of the body and enable the giraffe to keep its balance. The neck of a horse pulling a heavy load makes the same movements as the neck of a walking giraffe, and probably the necks of all large animals move in the same way, although the movements are not readily detected because the centre of gravity is farther from the forequarters than it is in a giraffe, thanks to its massive neck. In the giraffe the main propulsion for each forward movement comes more evidently from the forelegs than it does in other quadrupeds.

The neck plays an even greater role in a galloping giraffe than it does in a walking one, stretching farther forward and reaching farther backwards. When the legs of a galloping giraffe were bunched together under it during a stride, the neck was beginning to stretch forward (fig. 6.15B: the curve of neck–back angles against time was starting to rise). When the legs were spread out (D) the neck was stretched farthest forward. Therefore in the gallop as in the walk the neck arches backward and forward between each forward spurt, in each sequence the neck momentum slightly preceding that of the body. But whereas in the walk there were two cycles per stride, corresponding with the right and then the left legs moving forward, in the asymmetrical gallop there was only one. In either case the time for a neck cycle was similar, because a galloping stride took less than one second to complete, whereas a walking stride took about twice as long.

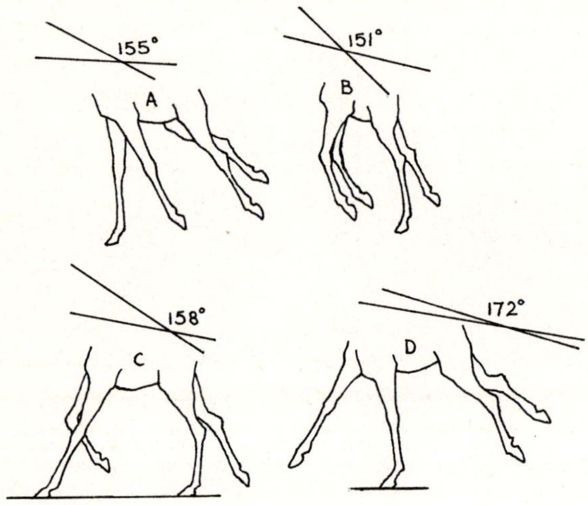

Fig. 6.15. Four phases in the gallop of a giraffe. The angles formed between the slope of the neck and the back are greatest when the giraffe's legs are spread out and least when they are bunched together under the trunk. After A. I. Dagg (1962), *Journal of Mammalogy*, **43**, 88–97.

7. Walking and Running in Lower Vertebrates

7.1. *Evolution and General Characteristics*

By the early Devonian, over 300 million years ago, plant life had become established on dry land; by the late Devonian, terrestrial vertebrates evolved from fish had followed the plants on to this potential wonderland. These amphibians were more fish-like than present-day amphibians in the structure of their bones and teeth, and especially in their tails. Unlike fish they had no fins, but four walking legs on which to move in the new medium. The forelimbs were used chiefly for pulling at first, and because of the heavy tail the hind legs chiefly for support.

There are two basic ways in which the locomotion of amphibians and reptiles with four legs differs from that of mammalian quadrupeds. Firstly, reptiles and amphibians, because they are poikilothermic (or, loosely-speaking, cold-blooded), cannot move quickly at low temperatures. Their output of power from a muscle is approximately doubled for every 10 K rise in temperature: at an ambient temperature of 10°C a mammal can develop eight to ten times as much locomotory power as a reptile or amphibian of the same size. V. B. Sukhanov of Russia found for example that he could not make a gecko run fast for filming unless he first carried it into bright light and sun-scorched the surface of the sand it was to run over. Secondly, with the possible exception of crocodiles, amphibians and reptiles do not gallop, nor have any asymmetrical gait. Thus their repertoire of gaits is simpler than that of mammals. Further, their walk is less variable, resembling a slow trot, so that it is often difficult to determine when this gait becomes a run; as they speed up, their leg movements quicken, but they do not change much otherwise. Some authors resolve this dilemma by defining a walk as a gait in which each foot is on the ground over 50 per cent of the time of a stride, and running as one in which each foot is on the ground less than 50 per cent of the time of a stride.

7.2. *Amphibians with Tails*

The locomotion of tailed amphibians is generally accepted as the primitive type for all terrestrial vertebrates. On dry land, salamanders

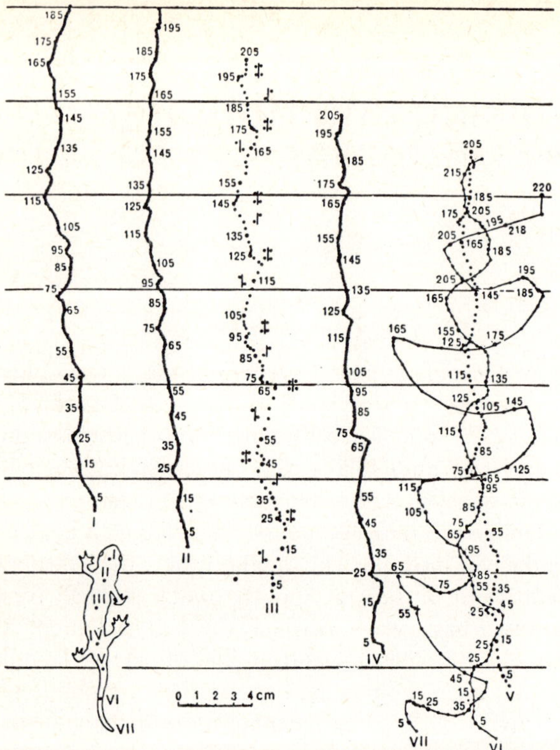

Fig. 7.1. Trajectories of dots painted on a salamander's back, during its trot-like walk. The supporting legs are indicated beside the movement of the mid-body dot (III). Numbers refer to film frame numbers. From V. B. Sukhanov (1974), *General System of Symmetrical Locomotion of Terrestrial Vertebrates*. New Delhi: Amerind Publishing.

and newts use symmetrical diagonally supported gaits already described for mammals, but with the upper parts of their legs in a horizontal plane (fig. 3.1). This posture requires energy to raise the body above the ground during walking or running, so between spurts of movement the animal usually rests its abdomen on the ground. During locomotion the hands and feet are oriented in the direction of motion, with the front and hind legs supplying equal propulsion. The body itself curves noticeably at each step so that each foreleg in turn can reach farther forward than it could if the body remained rigid. V. B. Sukhanov has filmed from above the walk of a spotted salamander (*Salamandra salamandra*) along whose back seven dots of white poster paint had been dabbed. Consecutive dots from each part of its body were then marked on paper as in fig. 7.1. When the dots of each series were joined, it was evident that the tail had the most lateral displacement during the walk, followed by the mid-body. These dots followed a rough sinusoidal pathway. The amplitude of the transverse variation was least in the pectoral and pelvic regions, because these were most closely attached to the ground via the limbs. Thus the lateral bends of the body during the salamander's walk were roughly stationary waves. The successive displacement of dots for the mid-body (III, not joined in fig. 7.1) indicate that movement of the salamander was not uniform, because the dots themselves were marked at the equal time intervals of every second frame. Where the dots are far apart, the dot (and therefore the mid-body) was moving fast; this occurred when the salamander was supported only by two diagonal legs. When the dots are close together, the body was almost stationary, and, as one might expect, the animal had all four feet on the ground. Thus during one complete stride the speed of the salamander's body increased twice and decreased twice.

Tailed amphibians walk as most mammals do, using their legs in the order left hind, left front, right hind, right front, slowly with only one leg off the ground at a time, or faster with diagonal legs supporting an individual during its strides. The two diagonal legs are often set down almost together in a gait approaching a walking trot. Thus their movements are as stable as possible, consistent with the speed at which the animal is going.

7.3. *Frogs and Toads*

The gait of walking frogs and toads is similar to that of tailed amphibians, with the exception that it is always slow and, because of their lack of tails and their compact bodies, there is little undulation of their bodies. The walking movements are cumbersome, but this is not important because they can hop or jump if they need to move quickly to avoid danger or to obtain food. The frog's jump is discussed in Chapter 8.

Various herpetologists have commented on the skittering locomotion of frogs in which they run over the surface of the water rather than dive into it. P. Chabanaud of Paris saw an African frog *Rana occipitalis* run for several metres on water, ricocheting along on its hind legs; he did not note if the legs were used alternately or together. R. Hudson, who watched cricket frogs (*Acris crepitans*) in Pennsylvania skittering over water after they had been disturbed from the shore, noted that after about a metre each dived into the water and swam rapidly back again to the bank, where many left the water to hide in the grass. He speculated that the fast movements of skittering and swimming may be made to avoid predators such as the bluegill sunfish (*Lepomis*) which readily eats these small frogs. H. Janson noted in North Carolina that green tree frogs (*Hyla cinerea*) often skittered over the water, but that they could only do so if they started the skittering with a leap begun well above the surface of the water. Those that he launched from his hands held near the surface did not skitter.

7.4. *Lizards*

Reptiles are more evolved than are amphibians for a terrestrial life, but their gaits are generally similar, consistent with their anatomy. Lizards, as the most generalized form of reptile, move their legs in the same sequence as tailed amphibians, which they superficially resemble. However, they tend to have relatively long hind and short front legs, so that the hind foot is often set down on the ground well ahead and to the side of the ipsilateral front foot (fig. 7.2). Because the legs are lateral to, rather than under, the trunk, with proximal segments which move in a horizontal rather than a vertical plane, the legs describe sweeping arcs as they move forward. Perhaps because of these movements the proportions of the bones in the elongated legs of cursorial lizards are quite different from those in the legs of cursorial mammals. As in the salamander, lizards have a great deal of horizontal flexion in their vertebral column which increases the length of each stride.

Reptiles can move faster than amphibians, changing from a slow walking trot to a true trot in which they are supported by only two diagonal legs. In films of the collared lizard *Crotaphytus collaris*, which has hind legs considerably longer than forelegs, R. C. Snyder of the

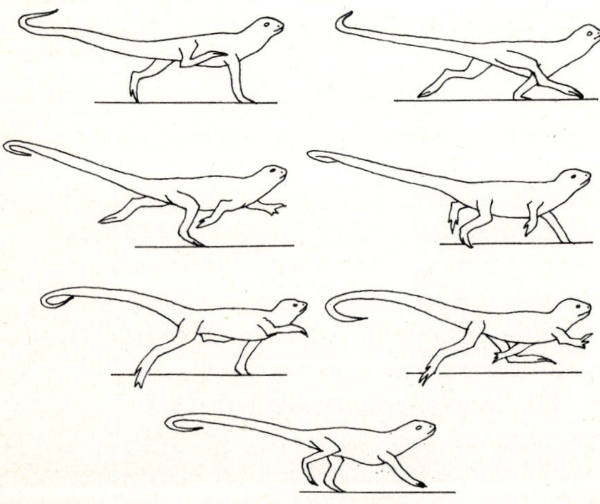

Fig. 7.2. Quadrupedal trot of the collared lizard. The short front legs touch the ground well after the contralateral hind legs. From R. Snyder (1952), *Copeia*, 64–70.

University of Washington, Seattle, noted that the length of stride was greater in the rear than it could be in the front, given the short front legs (fig. 7.2). Thus at a fast trot the front foot touched the ground later and left it earlier than did the opposite hind foot. At increasing speeds a lizard's trunk is held more vertically and higher above the ground, and the trunk undulates more with the longer steps, but the speed of each stride is increased only slightly.

If, during their evolution, lizards were selected for speed, one can imagine that the hind legs might increase in size and power to attain a longer stride and greater thrust, and the forelegs might cease to be used at high speeds. This is apparently what has happened. Over 37 species of lizards run fast by tucking in their forelegs, striding along on their hind legs whose step angles may reach almost 180°, and depending on their heavy tails to keep their balance (fig. 7.3). Thus bipedalism in lizards seems to have arisen from a symmetric trotting gait as a way to increase speed. In most mammals by contrast, speed beyond that attainable in the trot is achieved by switching to an asymmetrical gait, the gallop.

The widespread use of the bipedal gait in lizards does not mean that bipedal recent lizards and bipedal dinosaurs evolved from the same stock, as some zoologists have hypothesized, but that bipedalism is a relatively common phenomenon which has evolved in reptiles several times. This evolution seems closely correlated with powerful hind legs

Fig. 7.3. Bipedal locomotion of the basilisk. The hind legs describe a wide arc as they swing forward. From R. Snyder (1949), *Copeia*, 129–137.

in rapidly running groups. A bipedal run in a small animal is more economical than a quadrupedal one, because the length of stride of the hind legs can increase and the forelegs no longer retard forward movement by touching the ground. However, bipedalism in reptiles is inefficient because their non-resting posture, with their upper hind limbs extended out from the body rather than under it, necessitates strong musculature whenever the body is off the ground; thus lizards generally have large limb bones and muscles despite their small size. In large dinosaurs such a stance was probably impossible, especially if they moved bipedally, so the legs were rotated as close under the body as possible to enable them to bear the animal's weight directly.

Unlike man, in whom bipedalism has been accompanied by a complete 90° rotation of the trunk and associated drastic morphological changes (see Chapter 3), bipedalism in lizards has been accompanied by few anatomical changes. This is because the body remains in a more or less horizontal position whether the lizard is using four legs or two. In bipedal forms the front part of the trunk and the front legs tend to be shorter than in quadrupeds, the pelvis narrower, and the tail longer, but these changes could be a result of increased speed of running rather than of bipedalism *per se*. However, when the terminal one third of a lizard's tail was cut off, the animal could only run bipedally four or five strides without losing its balance and falling; when two thirds were removed, it could run no more than one step on two legs.

Bipedalism is found in lizards which live in warm regions of open sand and stones, in lizards which are semi-arboreal in bush country, and in lizards which are semi-aquatic near forest streams. Among the latter the basilisk (*Basiliscus*) uses a bipedal gait to skitter across the surface of water. Most lizards, especially small species, move remarkably quickly. Snyder's collared lizards were timed at $4 \cdot 9$ m s^{-1} quadrupedally and $7 \cdot 1$ m s^{-1} bipedally, and the teiid lizard *Ameiva ameiva* at $3 \cdot 0$ m s^{-1} and $7 \cdot 3$ m s^{-1}.

7.5. *Crocodilians*

Crocodiles and alligators have three types of movement on land, one slow and two fast. In their common slow gait the body is carried so close to the ground that it sometimes leaves a drag mark, as the tail does. The upper limb segments, like those of a lizard, move almost in a horizontal plane. This movement can degenerate into a belly run when the animal wants to slide into the water; then it drops completely on to its stomach and toboggans rapidly down a river bank, using its laterally spread legs as paddles.

The other crocodilian gaits are seen less often. In the fast walk the body is raised high so that the limbs are straighter, with the thigh at times perhaps 50° to the vertical plane, although the tip of the tail still drags along the ground. Here the limb action is much more pendulum-like than it is in the lizard, with most of the propulsive force taking place behind the hip joint. The foot is set down almost parallel to the direction of movement. During the walk the limbs act as diagonal pairs, as for the slow walking trot.

The third crocodilian gait, observed by G. Zug of the National Museum of Natural History in Washington, is the gallop. In this gait, which is found only in small animals up to two metres long, the hind and front pairs of legs each move more or less together, the latter for propulsion and the former for support, to attain a speed of up to 13 km h^{-1}. This is the only example known of an asymmetrical gait occurring in reptiles.

7.6. *Tortoises and Turtles* (*Chelonia*)

Locomotion in tortoises is somewhat affected by the encompassing shell, but not as much as one might expect. The typical reptilian limb position is a sprawled one, with the upper parts of the legs held parallel to the ground. The presence of a shell has not altered this basic posture, but the shell bridge has restricted the extent of the arc through which the limbs can move. It is possible that of all living reptiles the tortoises may possess a locomotor pattern closest to that of primitive forms, since the evolution of the shell early on in the history of this class may have fixed the original pattern of walking. However, it is probable that most early reptiles moved their axial skeleton to a considerable extent, an ability denied chelonians; axial locomotion is the primitive condition in vertebrate animals, and the dominant one in most aquatic forms. (The notable exceptions are chelonians and birds, both of which have rigid or fairly rigid trunk vertebrae which prohibit much body movement, and which therefore use their appendages in swimming.)

In order always to maintain their balance, terrestrial tortoises can only progress slowly, with each limb supporting the animal for at least 70 per cent of each stride. As fig. 7.4 indicates, the centre of gravity is always between the diagonally supporting legs (3 and 7 in fig. 7.4) or within the triangles made by the three or four supporting legs. A tortoise is usually supported on three legs 60 per cent or more of the time of the stride, and on two diagonal legs for a much shorter period. Tortoises, unlike most mammals, are never supported solely by their two left or two right legs; at their slow speed they could not balance on two ipsilateral legs, and each hind leg does not swing forward far enough for the leg in front of it to need to be moved before the hind can be set down. The sequence of limb movement is the same stable sequence found in most other quadrupeds (near hind, near front, far hind, far

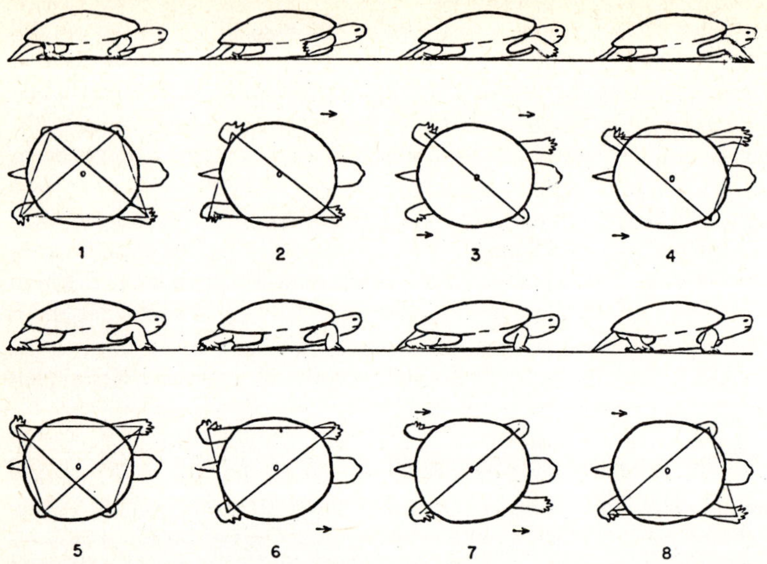

Fig. 7.4. Walk of the painted turtle. The centre of gravity at the middle of the shell always lies between the two supporting legs or within the triangle of three supporting legs. Arrows indicate which feet are moving. From W. F. Walker, Jr. (1971), *Journal of Morphology*, **134**, 195–213.

front), shading at relatively quick speeds into a slow trot in which a hind leg and its diagonally opposite front leg swing forward at the same time.

During progress the foot of the painted turtle (*Chrysemys*) is usually set down on the ground flat-footed, with the entire plantar surface and the tips of the claws striking the ground simultaneously. In those genera such as the snappers (*Chelydra*) and the soft-shelled turtles (*Trionyx*) with reduced plastra and a narrow shell bridge, the foot at footfall is anterior to the knee, thus increasing the length of the leg's step. In tortoises without these shell reductions the foot is planted under or behind, rather than ahead of, the knee. In *Chelydra* the leg can attain an angle of nearly 90°, in freshwater tortoises the step angle is not more than 80°, and for terrestrial tortoises the angle is about 65°. Depending on the species and its speed of progress, a tortoise may walk with its shell high above, or near to, or dragging on, the ground, and with its body always horizontal, or rolling and pitching during each stride. In these characteristics, and in movements of the legs, there is no easy correlation with the taxonomy of tortoises. The speed itself varies within a stride, being highest when only two feet are on the ground and lowest when four feet are there. Deceleration tends to occur when there is a shift from a bipedal to tripedal, or a tripedal to quadrupedal support pattern, while acceleration takes place when there is a shift to fewer supporting legs. If the plastron drags along the ground, as it tends to do in the semi-aquatic painted turtle, this also affects the speed.

Sea turtles may or may not move like their terrestrial relations. Loggerhead females (*Caretta caretta*), which crawl up sandy tropical beaches to lay their eggs well above the high-tide mark, do so using the quadrupedal gait observed in tortoises. When hatched, their tiny young run down the beach using the same alternate movements. Similarly, young and adults swim using their flippers alternately. By contrast, the female green turtle (*Chelone mydas*) uses pairs of flippers together in dragging itself along the sand. It moves all four flippers forward at once, then hoists its body along between them. Its body is too heavy to lift off the ground, so the trail it leaves is a broad trough, bisected along its length with a tail mark, and bordered at its edge by inner regular transverse furrows left by the hind flippers, and outer ones left by the larger front flippers. Its young move like their parent, pulling themselves forward with the front flippers acting synchronously. This species also swims using its pairs of flippers together, rather in the manner of the human butterfly stroke. With this stroke a green turtle can swim faster than 2 km h^{-1}.

7.7. *Dinosaurs*

Dinosaurs, which first appeared near the end of the Triassic 175 million years ago, radiated into a variety of evolutionary lines. The agile, bipedal group can be represented by *Struthiomimus* (fig. 7.5(*a*)). This animal had strong, bird-like hind legs; three-toed hind feet; small forelegs; a long neck; and a long tail which served to counter-balance the weight of the body which was pivoted at the hips. Unlike present-day reptiles, such bipedal forms had their legs rotated under the body in an efficient posture not evolved in mammals until relatively recently in forms such as man. Undoubtedly these animals were fast enough both to run from predators and to overtake prey which they seized with their teeth and front legs.

The most spectacular dinosaurs evolved into animals of great size. *Brontosaurus*, over 20 metres long, supported its vast weight on four legs, but the huge *Tyrannosaurus* (fig. 7.5(*b*)) seems to have moved on two legs: its tail did not leave a drag mark between its footprints as one might have expected in a quadruped; its forelegs were small; and the muscles and bones of each hind limb were apparently strong enough to support the full weight of the animal. Romantics have pictured *Tyrannosaurus* striding along majestically, but its gait was not like this at all. Rather, fossilized footprints show that its stride (0·65 m) was small compared to the length of its feet (0·3 m). It may have been that each femur was strong enough when vertical to support the immense weight of an individual, but not strong enough when held at an angle of 60° or so, the angle which the bone would attain during long strides. The footprints of a dinosaur of this species from Swanage as displayed in the Dinosaur Gallery of the British Museum (Natural History) lie in uneven line, with the toes pointing inward. Apparently *Tyrannosaurus* walked pigeon-toed, with a marked duck-like waddle.

Fig. 7.5. Representative fossil reptiles. (*a*) *Struthiomimus*; (*b*) *Tyrannosaurus*; (*c*) *Cynognathus*. All have their legs beneath rather than to the side of their trunks. Not drawn to scale.

Long before the first dinosaurs appeared, certain reptiles were evolving along lines that were to lead rapidly to warm-blooded mammals. By the Triassic these therapsids were mammal-like in many characteristics, including the important one of having not laterally sprawling legs as in present-day lizards, but legs which were placed directly under the animal. In *Cynognathus* they moved efficiently forward and backward, with little lateral swing (fig. 7.5(*c*)). Correlated with this advanced posture were compact and strong feet, well developed leg joints, and strongly interlocking and differentiated spinal vertebrae.

7.8. *Birds*

Birds evolved from thecodont archosaurs, reptiles which were evidently cursorial before becoming adapted to arboreal life. The avian ancestor was undoubtedly active rather than sluggish like most reptiles, and bipedal so that the front legs could evolve into wings. Like their ancestors, birds have cursorial legs judged by the proportions of their leg bones, with distal elements elongate compared to the femur (fig. 3.2(*d*)). Despite this, many birds can in fact scarcely walk. Seabirds such as albatrosses, shearwaters, and guillemots, which spend most of their lives on or over the oceans can only waddle when they come to land to reproduce. Because they are medium-sized or large birds, their legs are not strong enough to spring them high into the air so that their wings can take over the locomotion without hitting the ground as they flap; instead they must launch themselves into the wind to begin flying. Aerial birds such as nightjars, hummingbirds, swallows and martins virtually never walk; swifts are unable even to perch because their legs are mere stumps with four forward-pointing toes which can only cling to vertical walls. Most birds of prey also seldom walk because of their large curved claws; they perch on projections rather than on flat surfaces. Penguins are highly adapted for swimming in cold water by means of their much-modified wings, but their short legs which conserve heat and give streamlining make walking ungainly. They waddle with their flippers held out sideways for balance, proceeding with marked side-to-side swaying movements, and some species, especially going uphill, shuffle on their bellies, using their wings as well as their legs. Like the penguin, flightless birds are not necessarily cursorial, especially if they live on islands where they have no predators. Such birds include the marine steamer ducks (*Tachyeres*) of South America and the Falkland Islands; the Galapagos cormorant (*Nannopterum harrisi*); and two rails *Gallivallus australis* and *Notornis mantelli*, a parrot *Strigops habroptilis* and three kiwis (*Apteryx*) of New Zealand.

Flightless birds which have to contend with predators are nearly all large—ostrich, rhea, cassowary, and emu. They do not hop, but walk or run so efficiently that ostrich have been timed at 80 km h^{-1}. Because such birds have no tails for counterbalance, their femora slant forward from the acetabula, bringing the legs more nearly under the centre of gravity of the bird than they would otherwise be. When they run, some of them (such as ostrich and emu) use their wings as rudders which enable them to turn agilely; others (rhea and cassowary) hold their heads well ahead of them, either to avoid vegetation or to help their balance.

The birds with the relatively longest legs (cranes, herons, stilts) are not the fastest; they have become specialized for such things as wading in water, or counterbalancing a heavy head during flight, rather than for speed.

Many terrestrial birds hop, rather than walk. This is true of most small birds up to the size of the house sparrow, perhaps because (at a given speed) hopping does not require more power. It is also true of some low-dwelling forest or scrub birds such as tropical pittas (Pittidae), babblers (*Picathartes*), jays (*Garrulus glandarius* and *Cyanocitta cristata*), flickers (*Colaptes cafer*), woodpeckers (*Picus viridis*) and kookaburras (*Dacelo novaeguineae*). Gulls hop three or four times before take-off, and some large raptors hop on their rare visits to the ground. The American robin and blackbird (both *Turdus*) may either walk or hop (why one gait is used at any one time is not known), while the young of skylarks (*Alauda arvensis*) and young ravens (*Corvus corax*) hop even though the adults do not. The common mynah bird (*Acridotheres tristis*) of India sometimes hops briefly, then walks faster and faster until it breaks into an irregular two-step. The rockhopper penguin (*Eudyptes crestatus*) either walks when on shore or negotiates rocks in a series of hops with both feet together. That several species can both walk and hop indicates that the use of a gait is not a function of the muscular system, but of the nervous system.

Birds exhibit all types of terrestrial locomotion from the waddle of ducks, to the swaying gait of penguins which balance with their flippers held sideways, to the regal stalking of cranes. I. Clark and R. M. Alexander, who analysed the terrestrial walking of quail (*Coturnix coturnix*), found that the force records were very similar to those of human beings. As in human walking, each foot exerted a forward force on the ground followed by a backward one, so that the body was first decelerated and then accelerated. The vertical forces also showed two peaks, the first occurring when the foot first touched the ground and the second as the foot pushed off from the ground. In human beings however the vertical force is greatest while both feet are on the ground, while in the quail the total vertical force exerted by the feet while both were on the ground was less than the forces acting while only one foot was on the ground. The force records of the quail were larger relative to its weight than were the records of human beings.

Another difference between the walks of men and birds is the action of the main joint situated half-way down the part of the leg which is visible. The knee joint in man (femur–tibia/fibula) points forward, while the comparable tibiotarsus–tarsometatarsal joint in birds points backward, as shown for the reef heron (*Egretta sacra*) in fig. 7.6. The knee joint of a leg in a walking person (fig. 4.7) remains straighter

Fig. 7.6. Reef heron walking. The apparent 'knee' joint, tibiotarsus–tarso-metatarsus, points backwards not forwards as in man.

while it is acting as a support than it does in birds, although in neither is it uniformly straight. In the reef heron and the white ibis (*Threskiornis molucca*) the leg is rather bent when it is directly under the bird, and straightest as the leg touches the ground (i.e. highest hump follows the trough in fig. 7.7). In the white-headed stilt (*Himantopus himantopus*), magpie goose (*Anseranas semipalmata*), silver gull (*Larus novae-hollandiae*), and black oystercatcher (*Haematopus fuliginosus*), the leg is also bent in the middle of the stance phase, but straightest as the leg leaves the ground (i.e. highest hump precedes the trough in fig. 7.7). There is no comparable bending in the middle of the stance phase in ostrich, rhea, or emu. We shall not know if these differences are related to a bird's size, or to some other factor, until the walks of more species than those I have studied (mentioned above) have been analysed.

As with mammals, birds' gaits vary with the speed of the walk. The faster a pigeon goes, for example, the faster it bends and swings its legs forward; the speed with which it finally stops flexing its tibiotarsus–tarsometatarsal joint and sets its foot on the ground however does not change much.

Many birds, including the pigeon, thrust their heads back and forth in time with each step: the head is stretching out farthest as either leg begins its swing forward, and brought back farthest as the leg touches the ground again just as in the giraffe's walk. Thus the bird shifts its centre of gravity forward with each leg in turn, then helps to shift it back again as the swing ends (fig. 7.8). In the banded landrail (*Rallus phillippensis*) and the moorhen (*Gallinula chloropus*) the tail flicks up briefly as the neck stretches forward, which may also help to shift the

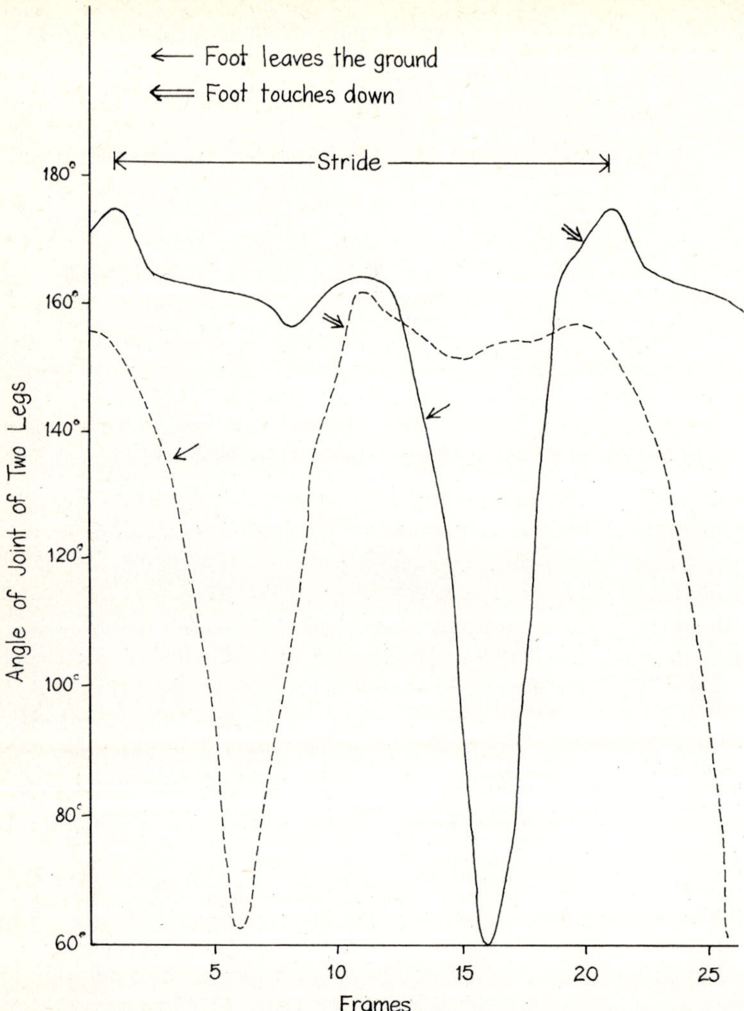

Fig. 7.7. Angle of the tibiotarsus–tarsometatarsal joints of the right and left legs of the white ibis as they swing forward during one stride. As in man, the legs are not uniformly straightened during the stance phase.

bird's weight during each step. The landrail and Australian magpie (*Gymnorhina tibicen*), and probably other birds too, also bob their heads when they are running, although to a lesser extent as their heads are already stretched well forward in this gait (fig. 7.9).

Birds that nod their heads back and forth as they walk change their total length about 15 per cent at each step. Some zoologists have

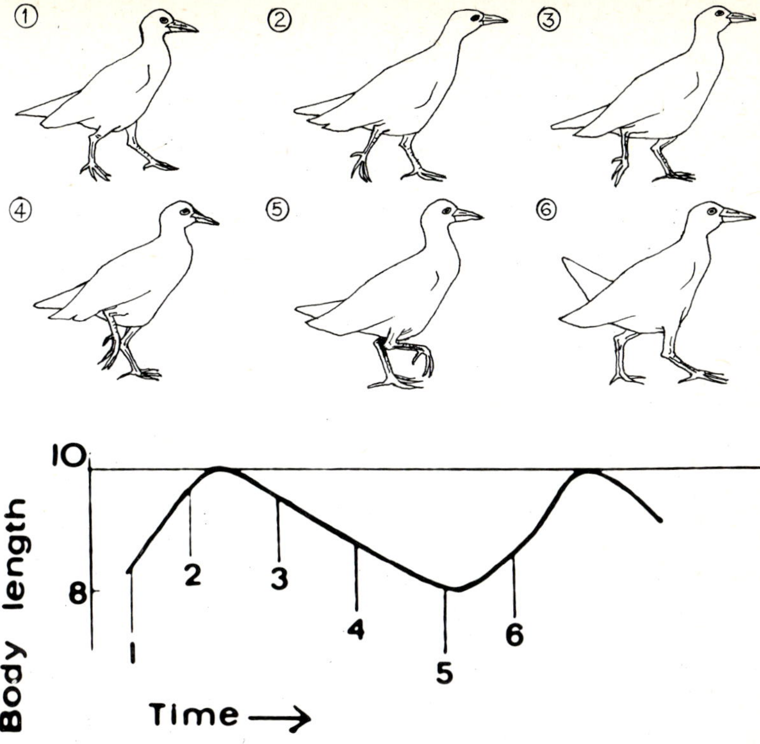

Fig. 7.8. Walking step of a landrail. The graph shows the change in body length because of neck movement during the step.

postulated that this movement enables a bird to see objects on the ground more clearly than it could if its head was moving forward uniformly. Others believe that it helps a bird to balance in the same way that a man swings his arms while walking to dampen oscillations about his centre of gravity, or a bipedal lizard uses its long tail. However, by tracing silhouettes of successive frames of birds such as the silver gull and the spur-winged plover (*Lobibyx novae-hollandiae*) I found to my surprise that these birds did not bob their necks at all as they walked. Nor do birds bigger than peafowl (*Pavo cristatus*). Why some species always use this head movement and some never do is a puzzle.

Very few studies have been made of gaits in birds as a taxonomic tool because so much ornithological energy has been expended in analysing their flight. One of these compared the walking gaits of four species of whistling duck (*Dendrocygna*). K. Rylander and E. Bolen of Texas found that the bending of the hip joint, knee joint, and ankle joint could all be correlated with the gaits of the different species. The angular

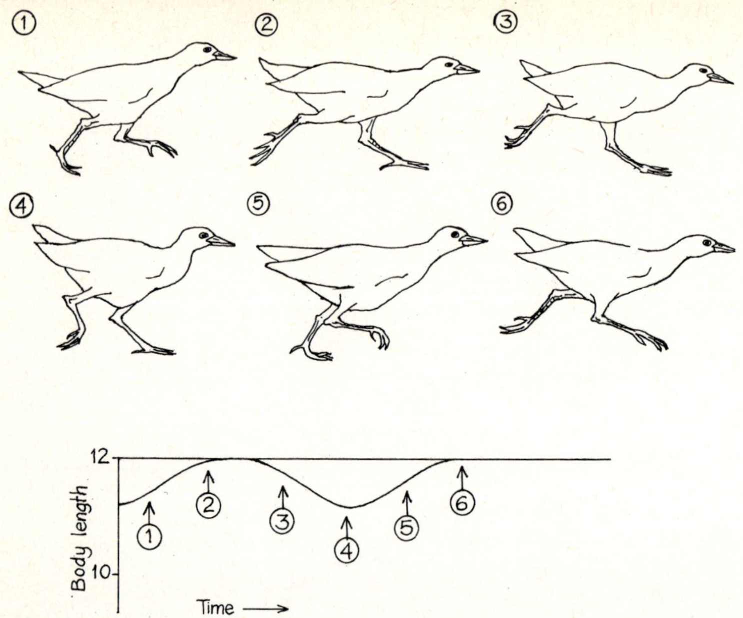

Fig. 7.9. Running step of a landrail. The graph shows the change in body length (to same scale as in fig. 7.8) because of neck movement during the step.

Fig. 7.10. Stance on land of four species of whistling duck. (*a*) *Dendrocygna autumnalis* and (*b*) *D. eytoni* have an upright posture and are relatively cursorial; (*c*) *D. bicolor* and (*d*) *D. arcuata* have a more horizontal posture and are highly aquatic. From M. K. Rylander and E. G. Bolen (1974). *Wilson Bulletin*, **86**, 237–245.

displacement was greatest at the hip joint and differentiated the two relatively cursorial species *D. autumnalis* and *D. eytoni*, which had a more vertical posture on land, from the two more aquatic ducks *D. bicolor* and *D. arcuata*, which stood more horizontally (fig. 7.10). In other characteristics the two cursorial species, one from the New World and one from Australia, were not especially similar, so perhaps their similar gaits evolved separately in response to their similar habitats. It seemed also in these ducks that the gait of the swimming species was the more primitive.

8. Hopping and Jumping in Vertebrates

8.1. *Filming in Zoos*

Although it is best to film sequences of moving animals in the wild, sometimes this is not possible. I found this to be true when I tried to photograph bush kangaroos in Australia. Although I could photograph the animals fairly well as they bounded among shrubs and past trees, it was impossible to separate their feet from vegetation and shadows when it came to analysing the individual frames from my film. For this reason I decided to base much of the material for a comparative study of kangaroo gaits on zoo specimens. In general this entailed much waiting around beside cages, because kangaroos of all species spend most of their time lying about. When I was feeling rich I hired a keeper to go into a kangaroo cage with me. (In one Californian zoo the rate for this was \$25 (£8) an hour in 1969, but keepers are essential; without one I once faced possible disaster while filming in a cage with an aggressive yak.) With the keeper to drive the animals past me from one end of the pen to the other, I was able to obtain excellent films of their locomotion. I could only hope that the animals had typical movements. Years before, I had filmed a sika deer at the Assiniboine Park Zoo in Winnipeg, Canada, which, on analysis, was found to have had a decided limp from an injured leg. Similarly I had many films of herds of Père David's deer (*Elaphurus davidianus*) and sitatunga (*Tragelaphus spekii*) in which all the animals were walking very slowly. Far from walking this way naturally, I found later that their slow movements were caused by the pain of stepping on cement. These species both have hooves which, when set down on the ground, spread apart widely so that they are buoyed up in the marsh areas among which they lived in the wild. This spread means that the flesh between the hooves can be rubbed raw to the point of bleeding on rough ground.

8.2. *Jumping, Hopping and Leaping*

Jumping and hopping are words which apply to a variety of movements, in all of which the centre of gravity of an animal is hoisted relatively high in the air. We should distinguish between a single jump or leap, in which an animal's momentum is caused by a single effort, and a series of jumps or hops.

By their very nature cursorial animals tend not to leap; they have evolved to move quickly over large distances, so a single jump would be of little use to them, except perhaps when they want to jump across a crevice or up a small cliff. Pronghorn and saiga antelope which lived on open steppelands a half-century ago were apparently unable to jump in the wild. Where fences had been erected by man on their home ranges, they crawled under rather than jumped over them. Presumably there was no physical reason for their refusal to jump the way other antelope do, but since they had never been confronted with obstacles they never learned or thought to jump them. Some antelope by contrast are excellent jumpers, although there is no obvious reason why they have developed this ability. Farmers in Africa who have decided to raise eland (*Taurotragus oryx*) as domestic stock have been amazed to find that these animals could leap over a two-metre fence without using a run before take-off.

Amongst the most effective leapers are prosimians, which use single jumps to move from one tree or branch to another. Tarsiers (*Tarsius*) can leap several metres with ease among the foliage, and bush-babies (*Galago*) can jump to a height of over two metres from a crouching start. Both these animals are small, with long hind legs to initiate a jump and nimble fingers to hang on to something firm at the end of it. Small rodents with enlarged hind legs can also give high single leaps, although they are more likely to include these in a series of smaller jumps used when trying to escape predation, as we shall see below. A kangaroo rat can jump over two metres if it is suddenly disturbed.

Animals which progress in a series of jumps can cover far more distance in one jump than they could from a standing start, just as man can jump much farther and higher with a running than from a standing start. The respective human records for the high jump are 2·50 m and 1·75 m, and for the long jump 8·90 m and 3·70 m. A greyhound travelling at high speed has covered over nine metres in one leap and a female red kangaroo (*Megaleia rufus*) 12·8 m. For a kangaroo being chased the highest jump recorded has been 3·2 m.

8.3. *Ricochet*

This potentially fast gait is a symmetrical one in which only the hind legs touch the ground in a sequence of jumps. Usually the legs contact the ground together as in the kangaroos (fig. 8.1), but in some species such as jerboas the hind legs can touch one after another, or indeed each contact may be followed by a period of suspension (fig. 8.2). Mammals which usually progress by hopping have enlarged hind legs and long and heavy or tufted tails which are held aloft at speed as a counterbalance. They can either be large like kangaroos, medium-sized like spring hares, or small like kangaroo mice or jerboas. Ricochetal terrestrial mammals are always primary consumers and usually live in desert areas.

Fig. 8.1. Sequence of a kangaroo hopping with support diagrams. From P. P. Gambaryan (1972), *How Mammals Run*. U.S.S.R.

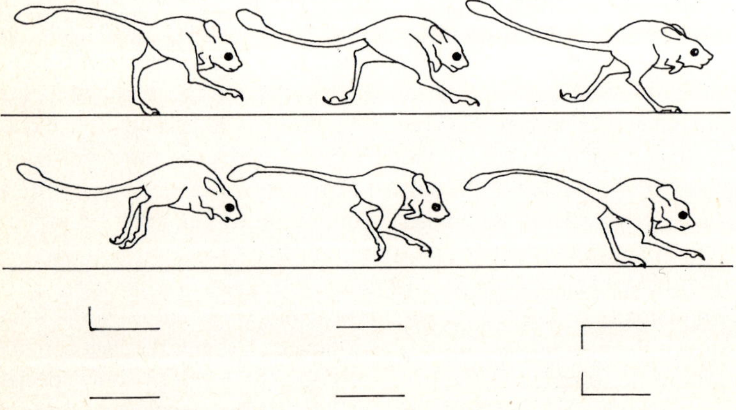

Fig. 8.2. Alternate ricochet of a jerboa, with support diagrams. The tail acts as a balance. From P. P. Gambaryan (1972), *How Mammals Run*. U.S.S.R.

Although they cannot be called ricochetal mammals, there are two other groups which hop on their hind legs on occasion. The first includes some lagomorphs, especially the northern arctic hare (*Lepus arcticus*) and Greenland hare (*Lepus groenlandicus*), which use it when they are closely pursued. They can climb hills and zig-zag readily with it, but its advantage over the half-bound is so far unknown. Perhaps the short tails of hares prevent a ricochet from becoming a major part of their locomotion. Leaps on the hind legs are also used occasionally in a series of successive spy-hops by white-tailed jackrabbits (*Lepus townsendii*) escaping in long grass. The second group includes various long-legged prosimians which leap from tree to tree in their arboreal habitat, but which also ricochet occasionally on the ground. They include the indris (*Indri indri*), woolly lemur (*Avahi laniger*), skifakas (*Propithecus*), bush-babies (*Galago*), tarsiers (*Tarsius*), and perhaps weasel lemurs (*Lepilemur*).

8.4. *Analysis of kangaroo jump*

A single jump of a kangaroo (*Macropus*) can be analysed as a mechanical motion. We can assume that the total mass of a kangaroo is concentrated at its centre of gravity (centre of mass) C, and that the animal leaves the ground at O with velocity v, at an angle α with the ground (fig. 8.3). During the jump gravity will act on the body giving a

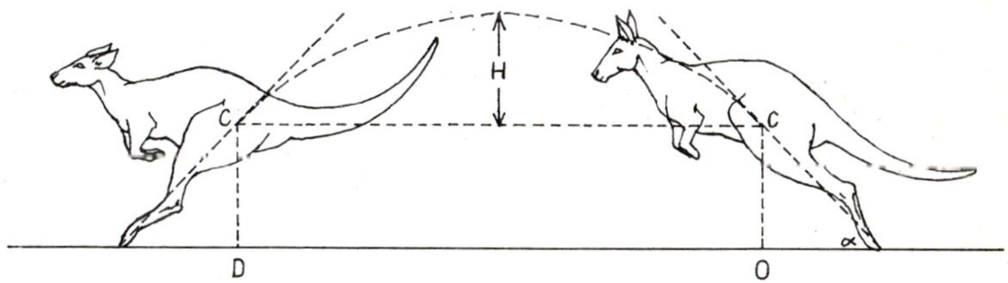

Fig. 8.3. Jump of a kangaroo. Its centre of gravity C moves a height H and a distance OD when it takes off at an angle α. After D. M. Badoux (1965), *Acta Anatomica* (S. Karger, AG, Basel), **62**, 418–433.

downward acceleration g of 9·8 m s^{-2}. This is a well known problem in physics which deals with the motion of a projectile under the action of gravity; if we neglect air resistance the path is parabolic. If we treat the kangaroo's flight in this way, we can predict that it will land with its centre of gravity just above the point D, where

$$\mathrm{OD} = v^2\,\frac{\sin 2\alpha}{g} \tag{1}$$

Because g has a constant value, the distance jumped depends on the starting velocity v and the angle α. This distance is maximum if α is 45°, since then $\sin 2\alpha = \sin 90° = 1$. Kangaroos, which have feet pads which give them a good grip on the ground, generally take off at about this angle of 45°.

We can also predict the greatest vertical height H through which the centre of mass C rose during the jump as

$$H = \frac{v^2 \sin^2 \alpha}{2g} \qquad (2)$$

Then, as one would expect, this greatest value of H occurs when $\sin^2 \alpha = 1$, thus $\sin \alpha = 1$, and $\alpha = 90°$. This takes place when all the animal's force is directed vertically downwards, so that it jumps straight upwards into the air.

The kangaroo's jump has been analysed by Dr. D. Badoux of Holland, who measured the distance OD as 6·1 metres in an animal jumping at an angle of about 45°. Thus from equation (1) we have $6·1 = v^2/g$, and the starting velocity of the kangaroo was

$$v = \sqrt{(6·1 \times 9·8)} = 7·73 \text{ m s}^{-1}.$$

This means that the animal is moving forward at a speed of 5·4 m s^{-1} or 19·4 km h^{-1} if we compute the horizontal component of its motion. The height H through which the centre of gravity rises under these conditions can be calculated from equation (2) to be 1·5 m.

It is also possible to calculate the upward force needed for a jump. From the moment at which the muscles of the hind leg begin to contract, to the moment its toes leave the ground, the kangaroo attains its upward speed. This speed is gained over the vertical distance between the position of the acetabulum in a sitting animal and that at the moment at which the animal leaves the ground. Again this can be treated as a problem in mechanics in which the work done by the animal (= *force on ground × distance*) is equal to the change in the kinetic energy. This vertical distance has been measured from photographs and the force exerted by a kangaroo jumping on the ground is estimated to be five times the weight of the animal.

In moving forward the kangaroo must also overcome air resistance. However, the energy needed to overcome this can be shown to be only about 1 per cent of the total expenditure of energy.

8.5. *Kangaroo Morphology and Evolution*

The kangaroo, like other ricochetal species, has an anatomy well adapted to jumping—a large tail as a counterbalance; short forelimbs;

long, sturdy hind feet; large hind quarters which shift the centre of gravity backward so that during hopping the animals can keep their trunks more or less vertical; a broad, sturdy pelvis to which muscles used in jumping are attached; hind limbs which move in a plane parallel to the trunk and close to it; and a resting position with the hind quarters considerably lower than the forequarters.

In kangaroos and wallabies the axis of the hind foot is through digit four, a circumstance unique among large cursorial mammals—perissodactyls have the axis through digit three, while artiodactyls and carnivores have it between digits three and four. The macropodids (large-footed ones) evolved from phalangerids, in which also the fourth digit was the main supporting one, and opposed the hallux, so that digits two and three were squeezed and reduced, and bound closely together except at their tips by a skin sheath, a condition called syndactyly. Thus there was no way in which the macropodids could evolve the use of the third digit in locomotion. In addition, whereas in ungulates the line of weight transfer by-passes the calcaneum, in macropodids this large bone plays an important part in transferring weight, as it commonly does in ricochetal placentals. Thus it may be that the evolution of the hind limb in kangaroos and wallabies, in which the fourth digit and calcaneum have been predominant, has effectively determined the gait which this family has adopted and perfected. Although at speeds of less than 18 km h^{-1} the hopping kangaroo expends more energy than a running quadruped of the same mass, at higher speeds it is cheaper energetically to ricochet. Perhaps in the Pleistocene many large Australian quadrupeds became extinct while kangaroos did not because of this.

The members of the kangaroo family have such a variety of gaits that it was possible for one of my graduate students, D. E. Windsor, to work out a phylogenetic tree for the Macropodinae using gaits as well as other taxonomic characteristics such as the shape of the molars and premolars, the anatomy of the reproductive system, the number of chromosomes, and the use of the tail which in *Dorcopsis*, a jungle species from New Guinea, is unusually flexible. All of the nineteen species studied possessed a slow progression, in which their weight was supported at times by the hind legs plus tail, and then the front legs plus tail, but only the tree kangaroos (*Dendrolagus*) possessed a normal walk similar to that found in most quadrupeds. These unusual animals live in trees, and although they usually hop or bound, the walk is used when they move slowly along horizontal branches (or ledges in the zoo). The tree kangaroos and quokkas (*Setonix*) are believed to be primitive because they possess a quadrupedal bound, with the two hind and the two front legs touching the ground alternately, without a period of suspension. This gait presumably evolved to a bipedal hop, the fast gait common to all kangaroos and the only one present in most of them. This ricochet has a relatively long period of suspension in those species such as the euro (*Macropus robustus*) and the black-faced kangaroo (*Macropus melanops*) which use jumps to negotiate rocks and heavy brush in their native habitat.

Large kangaroos can ricochet at speeds of up to 88 km h^{-1}, but most individuals have been clocked at maximum speeds of less than 50 km h^{-1}. It is impossible to know the extent of predation pressure on kangaroos in prehistoric times, first by now-extinct predators and then by dingos, but it may not have been great. Kangaroos are non-migratory and when they are stampeded they often pause to look back to see what startled them; it appears that during their evolution kangaroos were not selected because they had great speed, but because of other attributes such as the ability to breed throughout the year and to delay implantation of an embryo.

8.6. *Jumping Small Mammals*

Large animals in the wild are often hard to film, but small animals in the laboratory are even more difficult. It is relatively easy to catch them and build a glass run for them with back mirrors and grid markings, but it is frustrating waiting for them to use the gait you wish to record. Some refuse to move at all, and others never move smoothly, preferring alternately to sit still and to leap about nervously. When they do move they move so quickly that you may be unable to follow them with the camera, and if you photograph from a distance to ensure that all their leaps are on film, the animals appear very small. Zoologists in California have had good luck with small jumping rodents, especially kangaroo rats (*Dipodomys*). These animals are admirably named because their gaits are almost the same as those of kangaroos—slow progression using the tail, and bipedal hopping. In addition, when frightened they have high erratic leaps, which would be relatively physically impossible for heavy kangaroos to emulate. They can also walk bipedally, using one leg after the other, as true kangaroos are incapable of doing. Kangaroo rats have tails relatively even longer than those of kangaroos, which give their movements remarkable control (fig. 8.4). G. Bartholomew and H. Caswell watched one in the desert leap and land just beyond the entrance to a burrow which was at right angles to its line of flight and about 15 cm to the left. Apparently the animal saw this refuge as it passed, because without hesitation it leaped again, making a more than 90° turn and diving into the hole which was barely large enough for its body. The tail serves to balance a ricocheting kangaroo rat both by its position and by its active movements. When one individual's tail was cut off, the animal still jumped well at slow and medium speeds, but when it attempted a high leap it turned a complete forward somersault in the air and landed off balance. All leaps have to be high enough to allow the animal time to swing its hind legs far forward for the landing, and it may be that the long pes is an adaptation to extend further forward the reach of the hind feet at touchdown. *Dipodomys* has evolved a gait not of high speed or enduring locomotion, but of regular or erratic jumps which allow it to forage perhaps 20 metres from its burrow and to avoid predators while navigating in its desert habitat. Its forelegs are not essential to its locomotion, but are important in gathering food.

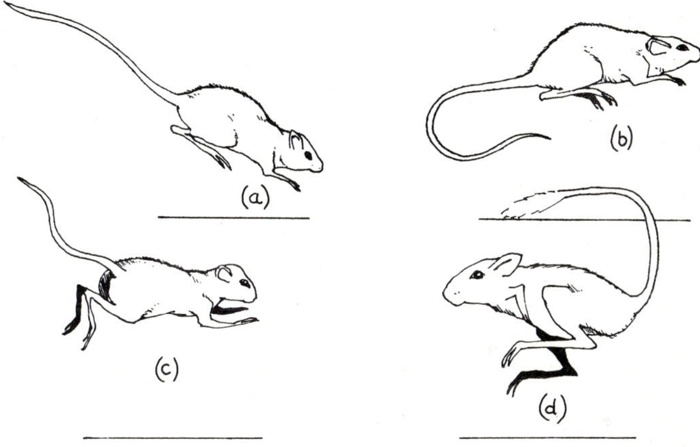

Fig. 8.4. Jumping of small rodents. (*a*), (*b*) and (*c*) Pocket mice; (*d*) Kangaroo rat. In both genera the tails are used extensively for balance during jumping. Based on the work of G. A. Bartholomew and G. R. Cary (1954), *Journal of Mammalogy*, **35**, 386–392; and G. A. Bartholomew and H. H. Caswell (1951), *Journal of Mammalogy*, **32**, 155–169.

The locomotion of the less specialized pocket mice (*Perognathus* spp) is interesting because although these desert rodents also have enlarged hind legs, reduced front legs, and long tails, unlike the related kangaroo rats but like kangaroo mice, they never hop bipedally. Instead their long leaps are incorporated in a bound, with push-off from the hind feet and landing on the front feet. Before the next leap the hind legs swing far forward, overreaching the point of contact of the front feet. If a pocket mouse is frightened it can leap a number of times in succession so erratically that one cannot follow it by eye. The leaps, which can be over 0·6 m high and 0·9 m long, are of course controlled largely by the long tail. Pocket mice assume a bipedal position when sifting through the sand and soil for food, but because their forelegs are short their bodies assume a horizontal position while they do this.

Kangaroo rats and pocket mice have similar anatomical adaptations, eat the same types of food, and live successfully in the same habitat. Then why is one bipedal and the other quadrupedal? Some zoologists have suggested that the pocket mice may be evolving in the direction of bipedalism, and that when during fast movement the overstep of the hind legs overreaches the front feet to such an extent that the contact of the front feet impede the forward momentum, bipedalism will become pre-eminent. Or possibly large hind feet have developed in both genera because the animals can then balance on them readily while searching for food with their front legs. In that case the method of locomotion used by the hind legs could be of secondary importance; as long as both genera can make strong, erratic jumps with their hind legs, as they can, whether they use forelegs or not in their fast movements may not matter.

Australia has another example, described by B. J. Marlow of the Australian Museum of Sydney, of two small animals which live in the same desert habitat but which move differently. Both the jerboa marsupial mouse *Antechinomys spenceri* and the rodent *Notomys cervinus* have long hind legs and long tails, but *Antechinomys* (which has the longer front legs) always moves on four legs as do the pocket mice, while *Notomys* can run both quadrupedally (at slow speeds) and bipedally (at fast speeds) as do the kangaroo rats (fig. 8.5). When

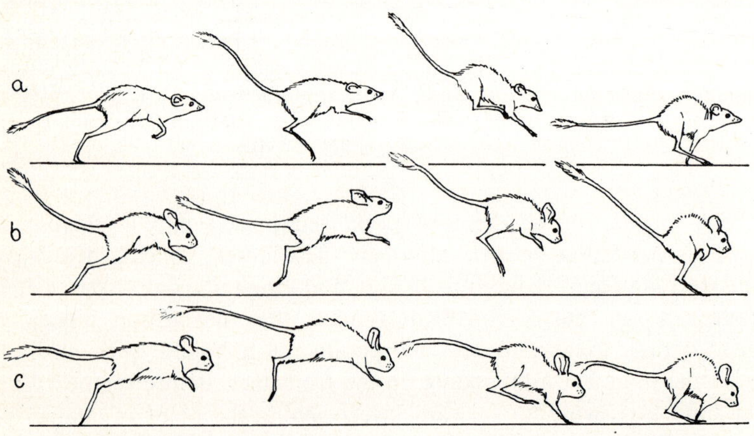

Fig. 8.5. Gaits of *Antechinomys* and *Notomys*. *Antechinomys* runs quadrupedally (*a*) while *Notomys* runs using either two legs (*b*) or four (*c*). Based on the work of B. J. Marlow (1969), *Journal of Zoology* (London), **157**, 159–167.

the front legs of *Antechinomys* were bound temporarily to its thorax with several turns of knitting wool, it was not only unable to run, but unable to stand, since it fell onto its side as soon as it was placed on the ground. Marlow found that the maximum speed of the bipedal *Notomys* was slightly faster (4·0 m s^{-1}) than that of the quadrupedal *Antechinomys* (3·6 m s^{-1}) although both had maximum leaps of nearly 0·7 m. In both animals moving at full speed the tail is used as a counterpoise for balancing, reaching its maximum height on landing to act as an air brake in reducing speed, and streaming out backwards on take-off. It is of much less use in *Notomys* when it runs on four legs, which is for it a slower and stable gait. Whereas *Notomys* is a prey species requiring swift evasive action in escape, *Antechinomys* is a predator feeding on invertebrates. Perhaps *Antechinomys* needs agility in catching and killing spiders and centipedes without itself being bitten. In any case both animals are well adapted to existing in desert areas despite their different modes of locomotion.

Experiments have shown that the rodent *Notomys* hopping quickly on two legs uses less energy than would a quadruped of the same mass moving on four. Yet hopping has not evolved as a gait for cursorial birds, which are all alternators, perhaps because it is not as fast as running. Probably the ricochet evolved in rodents for reasons other than speed. In small animals the ricochet is a useful gait for accelerating quickly from rest, for moving easily over irregular terrain, and for allowing erratic direction changes which confuse predators. Such changes are relatively easy when jumping because both legs are available to push upward at an instant, and if necessary at an angle too, although large upward thrusts use up a great deal of energy in raising the centre of gravity. By contrast alternators which do not give erratic high leaps can use more energy in forward propulsion and thus in increasing speed. In comparing the types of movements it is evident that quadrupeds do not have to worry about balance, as bipeds do; that bipedal alternators move with a greater conservation of energy than saltators which include high jumps in their locomotion do; and that such small saltators are interested less in conserving energy than in surviving in open terrain by moving with irregular leaps.

8.7. *The Jump of Frogs*

As we saw in Chapter 7, frogs walk by moving their legs alternately and more or less diagonally, their bodies stretching on one side and then the other, but their jump or hop is entirely unrelated to this type of progression. In the jump a frog always starts from a sitting position and is propelled upward by both hind legs acting together. The end of each jump does not give the animal any bounce for another jump, as the jump of a kangaroo does; a hopping frog is rather engaged in a series of standing high jumps. Correlated with this difference in dynamics are differences in usage. The jump of the frog is advantageous because it enables the animal to move fast and erratically in an unpredictable direction, to jump over dense vegetation, to leave a discontinuous trail which will be difficult for a predator hunting by scent to follow, and to confuse sight-predators which hunt by following movements of their prey.

In jumping, a single backward push must lead to a considerable amount of kinetic energy, because there is no further chance to add to this once an animal is launched into the air. Thus the mechanics of the jump are far simpler than of other forms of progression, as we saw for the kangaroo. The average force necessary for a jump can be calculated by the equation $F = ma$, where F is the average force exerted at the ground, m is the mass of the animal, and a is its average acceleration. This average acceleration a is $(v_1 - v_0)/t$, where v_1 is the take-off velocity, v_0 is the initial velocity (= 0 when the jump begins from a non-moving start) and t is the time during which the force is exerted, so $F = m(v_1 - v_0)/t$. In a jumping frog the force is supplied by contraction of the leg muscles and is the limiting factor. For the increase in velocity $(v_1 - v_0)$ to be large the time t must also be large.

Jumping animals, including frogs, grasshoppers, and kangaroos, start the jump in a sitting or squatting position with several sets of long leg segments folded against each other. During a push-off these segments unfold more or less at the same time, and are almost completely unfolded by the time the toes leave the ground. The longer the legs or the larger the number of folds, the farther the distance and the longer the time there will be through which the body can be moved while the foot is still touching the ground and can exert force through it. As well, the lighter the body, and the lighter the distal segments of the legs, the

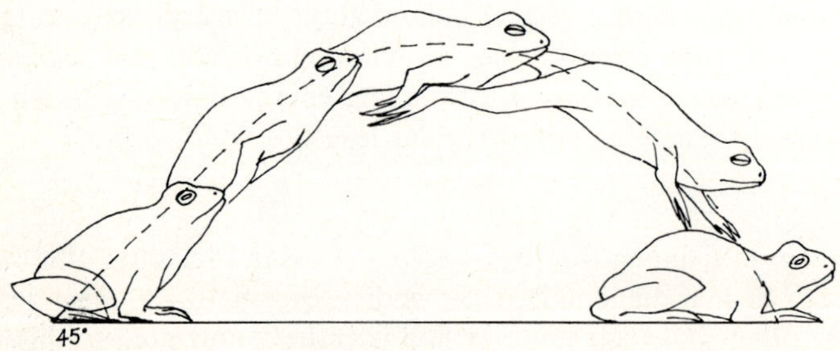

Fig. 8.6. Jump of a frog. The frog takes off at about an angle of 45° to the ground, pushing off for as long as possible with its long legs and landing with a splat.

smaller is the force required to provide the kinetic energy for the jump. The jumping efficiency is increased by the streamlining of the frog's body during the jump: C. Gans found from photographing a bull frog that its hind legs stretch out nearly straight behind it, its forearms are folded back against the body, and its eyes are closed and retracted into the head, the latter an adaptation which will also protect the eyes in the event of a collision (fig. 8.6). The forearms are stretched forward towards the end of the jump to help the frog land.

The flight of a jumping frog's body can be simply detailed, if one ignores air resistance, by calculating the speed and direction at which it is travelling just after the hind feet leave the ground, just as the path of a projectile can be calculated. This was done earlier for the kangaroo, and the same equations apply. That is

$$OD = \frac{v^2 \sin 2\alpha}{g}$$

where OD is the length of the jump, v is the initial velocity, α is the angle to the horizontal at which the frog moves initially, and g is the acceleration due to gravity ($\simeq 9 \cdot 8$ m s^{-2}). The length of the jump is thus not directly a function of the frog's size. The frog reaches its highest displacement H when

$$H = \frac{v^2 \sin^2 \alpha}{g}$$

The durations of upward and downward phases are theoretically equal and the path of flight symmetrical about a vertical line drawn through the high point reached by the frog's centre of gravity. Such a symmetrical path was acceptable for the kangaroo, an animal so large that air resistance was indeed negligible, but is much less so for the frog whose flight is noticeably affected by air resistance. The horizontal velocity does not remain constant throughout the jump, but decreases continuously. In fig. 8.6 it can be seen that the frog goes further in a horizontal direction on the way up than on the way down, in other words the upward phase is greater than the downward phase. The vertical component of the velocity is also decreased by air resistance, so that the average speed on the way down is less than the average speed on the way up, and the downward phase lasts longer. The maximum range of the jump neglecting air resistance occurs when $\alpha = 45°$. At this angle half of the energy supplied by the muscles is used to lift the body against gravity, while the other half provides the frog with forward horizontal momentum. If the angle of take-off is reduced to $40°$ or increased to $50°$, or changed to $30°$ or $60°$, the length of the jump is only reduced by about 2 per cent and 14 per cent respectively. Thus the speed at take-off is much more important than the angle of take-off in determining the length of a jump. Record jumping frogs include a bull frog (*Rana catesbeiana*) which jumped nearly five metres in three successive jumps, and a 50 mm South African frog *Rana oxyrhynchus* which jumped nearly 10 metres in a single leap.

Locomotory ability in anurans is closely related to systematics, particularly at the familial level, but also in lower levels of classification where limb proportions appear commonly in specific and generic keys. G. Zug of the National Museum of Natural History in Washington has studied the locomotion and bone structure of three types of anurans: the southern toad (*Bufo terrestris*), the green frog (*Rana clamitans*), and the tree frog *Hyla crucifer*. The short-legged toad had the least jumping ability, the frog the greatest absolute ability, and the tree frog the greatest relative jumping ability, when the length of the frog as well as its jump was considered. Zug noted that the strong jumpers tended to land with a splat, with the front legs absorbing the initial landing force but the chest and abdomen still hitting the ground forcefully. This landing left the hind legs sprawled beside rather than under the frog so that they had to be drawn closer to its body before the next jump. In contrast the toads landed gracefully after their shorter jumps without needing to readjust their posture. The strong jumpers have relatively short forelimbs, perhaps reflecting the need for stockier segments to absorb the shock of landing without breaking; large front feet to spread the touchdown force over a large area, thus reducing the force per unit area; a short scapula to reduce the amount of landing shock transmitted from the pectoral girdle to the vertebral column; and relatively long hind limbs and tibiofibulas which increase the length of jump.

Zoologists have wondered how the symmetrical jumping gait of the frog evolved from the basic quadrupedal movement of amphibians. Some have hypothesized that it evolved in an aquatic ancestor as a mechanism for driving the body forward in a straight line, thus enabling the animal to burrow readily into mud or vegetation. However, many fish and salamanders burrow easily using an undulatory system of locomotion. Others believe that jumping evolved in terrestrial pre-frogs as a quick method for returning to water by an animal relatively unsafe on land; frogs' eyes cannot see very well, but they may be able to direct the animal toward the nearest open areas. At first the jump could be completely unorganized—long or short, efficient or inefficient—as long as it carried the pre-frog to the safety of water. Once the simultaneous movement of the legs had been achieved, there must have been a rapid change in the body of the ancestral frog; disappearance of the tail which was no longer needed for propulsion or steering; shortening of the body so that the thrust of the hind legs had more impact; and streamlining of the head to counteract in part the shortening of the trunk. Because the propulsion of the swimming frog now came from the hind legs, the digits of these became webbed to increase their forward thrust. Still other zoologists feel that the trend towards a shorter body and stronger hind limbs was an aquatic one. None of these theories can be accepted absolutely until more fossils are found to enable palaeontologists to fill in the many gaps in the phylogeny of the anurans.

8.8. *Jumping over Obstacles*

Whenever a quadruped jumps over an obstacle, it does so with its head stretched forward, both to see where it is going and to keep its centre of gravity as far forward as possible. As it completes a jump, it raises its head, thus slowing its speed down somewhat. With the idea of a horse's changing centre of gravity in mind, some horsemen ride their horses leaning well forward as they start over a jump and well back as they complete it, but others try rather to remain as still as possible so their mount may not be unbalanced. Neither type of rider can overcome the fact that the horse and its plains ancestors were never built for jumping; with 18 pairs of ribs (compared to 13 in greyhounds) there is little room for flexibility between the rib cage and the pelvis.

The function of the neck and head in a jumping animal can be analysed most readily in the giraffe. Fig. 8.7 shows pictures from a film sequence of a bull giraffe jumping over a 1·5 m fence, together with the angles which the line of the back made with the slope of the neck. First it looked down at the fence (A), swung its neck far back and nearly perpendicular to the ground so that its weight was over its hind legs, and then hopped its two front legs over the top wire (B–F). Then it moved forward with the wire under its belly (G), swung its neck far forward so that its centre of gravity was over its forelegs, and hopped its hind legs over the wire together (H–J). This manoeuvre took less than one second. Jumping fences is an accomplishment which giraffe have only learned recently. When fences were first erected on what had previously been wild bushveld in South Africa they ran right through them, dragging wires and posts as they went. It took several years for giraffe here as well as for those in Kenya to learn to go over instead of through these obstructions.

8.9. *Jumping on to an Object*

Recently R. M. Alexander of the University of Leeds has studied the mechanics of jumping in the dog using a 36 kg male Alsatian called Happy in his tests, first in a long jump in which the dog jumped

Fig. 8.7. Sequence of a giraffe jumping a 1·5 m fence. The angles between the line of the back and the slope of the neck are given. From A. I. Dagg (1962), *Journal of Mammalogy*, **43**, 88–97.

3 metres over a series of low hurdles each 0·1–0·2 m high, and then in a high scale jump in which from a sitting start the dog leaped up and scrambled over a 1·88 m high wall. In the scale jump a vertical force equal to Happy's weight was recorded as he sat on the force platform. Then he straightened his front legs, lifted his forequarters into the air, and extended his hind legs until they too left the platform. The greatest force was produced by Happy as he pushed off with his front legs, and the next greatest as he pushed off with his hind legs. (These forces can be felt if a dog one is holding on one's lap decides to jump onto a nearby table). The two peaks of vertical forces occurred 0·35 s apart, and the dog's take-off velocity was about 5·5 m s^{-1}. In the high jump the dog's trunk was held much more steeply than in the long jump because the knees were more strongly bent.

The peak values of the stresses which acted in the major groups of hind limb muscles, and in the triceps were $190–310 \times 10^3$ N m^{-2} for the long jump and $230–260 \times 10^3$ N m^{-2} for the scale jump. Tensile stresses up to $60–80 \times 10^6$ N m^{-2} and compressive stresses up to 100×10^6 N m^{-2} acted in the tibia and humerus during the take-off for the jumps.

9. Tracks of Vertebrates

9.1. *Identification of Tracks*

Prehistoric men, as hunters, must have studied the tracks of animals for many thousands of years, just as nomadic people today still study the prints of human beings, camels, or horses. To an Australian Aborigine, for example, a person's tracks are almost as distinctive as his physical appearance or his voice. He is pleased to see a friend's footprints which are deep, with well-rounded heelprints, because these characteristics show that the friend is in good health. If he studies the edge of these tracks, he will also know how recently his friend has passed by, and what the local weather has been like since that time. The under-surface of a camel's foot is lined just as a man's sole is, so that Saharan camel owners can distinguish individual camels by their tracks. A man may trail a lost camel for many kilometres to bring it back to his herd. Mongolian nomads are as devoted to their horses as Saharan nomads are to their camels, and can distinguish up to ten different gaits in a horse, most of which are also apparent in its tracks. These tracks reveal how comfortable a horse would be to ride, and how sure-footed it is on uneven ground.

Most naturalists who study tracks are not interested in identifying individuals, but rather species. City people are often amazed to realize from tracks in the snow that rabbits or foxes regularly visit their gardens at night. In the desert, on beaches, near mud, or where there is snow, zoologists can make a fairly accurate record of what animals live in the area without ever seeing them. Sand is ideal for tracks in the early morning when dew gives its moisture; winter trails are best after a fresh light fall of snow. If footprints themselves cannot be identified for sure, their layout may be important. Thus Ernest Thompson Seton found that wolves and dogs could not be distinguished for certain by their prints, but they usually could be by their trails. Those of dogs were direct and unafraid, approaching buildings and farm equipment without hesitation; those of wolves were cautious, skirting around open spaces and areas frequented by man.

Animal tracks can give an excellent idea of what an individual was doing when it passed by—perhaps stopping to listen, perhaps pausing at intervals to sniff about for food, perhaps lying down to rest. They can also tell what gait an animal was using. In fig. 9.1 the walk and the trot are obviously the slower (because shorter) symmetrical gaits, and the two gallops the faster, asymmetrical gaits. The faster the gallop, and the more spread out the hoofprints, the more likely there is to have been a period of suspension after the front feet left the ground. The prints themselves may also reveal the speed of an animal. If they are deep, the animal landed on them heavily and probably while moving fast; if they are more or less shallow, depending on the nature and moistness of the substrate, the animal was going slowly. Tracks show which animals use a perfect tread, as cats do, putting the hind foot on the mark left by the front foot on the same side. This type of movement saves energy in snow, where one leg instead of two can do the work of

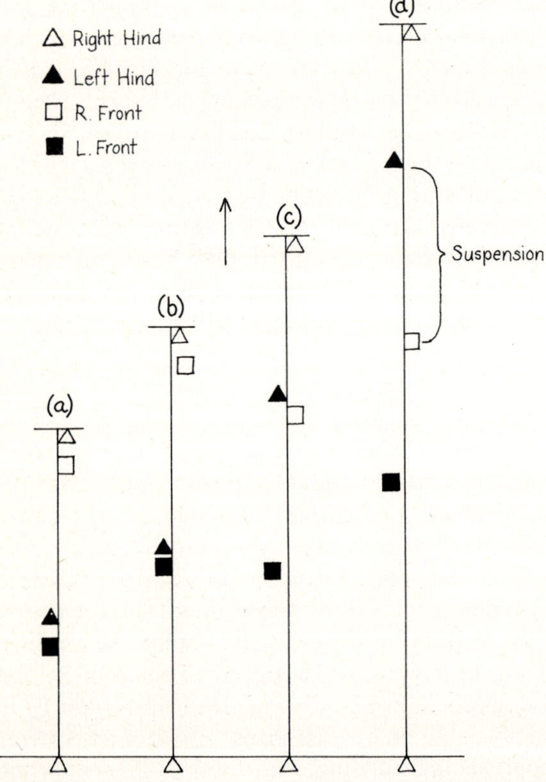

Fig. 9.1. Hoofprints of one stride in the gaits of the horse. (*a*) Walk; (*b*) Trot; (*c*) Slow gallop; (*d*) Fast gallop. The strides become longer the faster the speed.

shifting snow aside and stamping it down. It is also quieter, because once the front foot has found a footing without noise, the hind foot can follow it exactly. Dogs are noisier than cats because they do not tread perfectly and often drag their toes. In marshy areas there is an advantage in heavy animals not perfect-treading, because then their feet are less likely to become deeply mired in mud.

9.2. *Classification of Gaits*

Despite the millenia over which hunters, and the centuries through which naturalists, have pondered over animal tracks, no attempt has been made until recently to classify them by the gait used to produce them. The pioneer in this work was the Russian P. P. Gambaryan, who produced the key given in table 9.1 and illustrated in fig. 9.2.

The main difficulty in analysing tracks so that they tell something about the gait of an animal is that even if all the impressions of the feet can be readily seen and measured, it is usually impossible to determine which foot made each mark. This is true even if one watches an animal actually making the marks, as can be seen by trying it on a dog or cat or horse. Without such information, one can be seriously misled about the nature of the gait that was used when the track was laid down. In the simple set of tracks of a slowly galloping horse set out in fig. 9.3(*a*), there are three erroneous ways in which the gait could be interpreted (fig. 9.3(*b*)–(*d*)), in each case from right to left the two hind and two front hoofprints of a stride being grouped together. Fig. 9.3(*e*) shows the way in which the track was actually produced. In this example the direction in which the horse was going was known. Sometimes it is even impossible to determine this, if wind or rain have badly weathered a track.

9.3. *Records of Tracks*

There are four ways in which to make records of prints or tracks. One is to cast a footprint in a mixture of plaster of Paris and water. Such casts are excellent if made from a print in mud or ice, but poor if made in snow or dry sand. Their use is important in establishing the presence of a species in an area where it is usually not found. Casts have been made of what appeared to be cougar prints in north-eastern United States, and wildcat prints in northern England. The possibility of proving that either of these cats had wandered outside its usual range is greatly strengthened if a plaster cast has been made from which careful measurements can be taken.

Another method of making tracks is to dip the feet of an animal in different colours of paint and induce it to walk or run over a long strip of white paper. This procedure is far more difficult than it sounds.

TABLE 9.1. *Key to gaits according to tracks. After Gambaryan, 1972.*
To deduce the gait used to make a track, choose between the ordering number
on the left and the bracketed number.

		Examples
1 (6).	Only tracks of hind feet present.	
2 (3).	Footfalls distributed successively at an equal distance from each other . . . symmetrical walk and run (fig. 9.2.1).	Ostrich
3 (2).	Footfalls not distributed at an equal distance from each other.	
4 (5).	Footfalls distributed consecutively . . . half-paired and alternate ricochets (fig. 9.2.2).	Jerboa
5 (4).	Footfalls distributed in pairs . . . paired ricochet (fig. 9.2.3).	Kangaroo
6 (1).	Tracks of both hind and forefeet present.	
7 (8).	Footfalls alternate in the following order: hind-fore, hind-fore, or the hind footprint covers the fore footprint; the distance between the tracks on one side is always even . . . symmetrical gaits (fig. 9.2.4–6).	Basset, Giraffe, Cow
8 (7).	Footfalls alternate in the following order: hind-hind, fore-fore, or are distributed in pairs: hind feet-forefeet; if the tracks are distributed in another order, they are grouped in fours and each group is at a considerable distance from the preceding one.	
9 (12).	Tracks of all feet distributed consecutively.	
10 (11).	Footfalls of the right and left sides alternate . . . diagonal gallop (fig. 9.2.7).	Horse
11 (10).	Footfalls of the right and left sides do not alternate . . . lateral gallop (fig. 9.2.8).	Cheetah
12 (9).	The tracks of at least one pair of feet are set in pairs.	
13 (16).	Footfalls of the hind limbs paired, of the fore-limbs consecutive.	
14 (15).	As a rule, the increase in the distance between the hind and the fore footfalls goes along with an increase in the distance between the tracks of the fore and hind feet . . . half-bound (fig. 9.2.10).	Rabbit
16 (13).	Footfalls of both the hind and the fore-limbs paired . . . bound (fig. 9.2.11).	Quokka

Even the most phlegmatic pet is inclined, during the experiment, to
race off the paper, or rush back and forth in one spot, or leap over the
paper entirely, or sit down and lick its paws. The chances of obtaining
tracks of a wild animal in this way are virtually nil, except by accident.

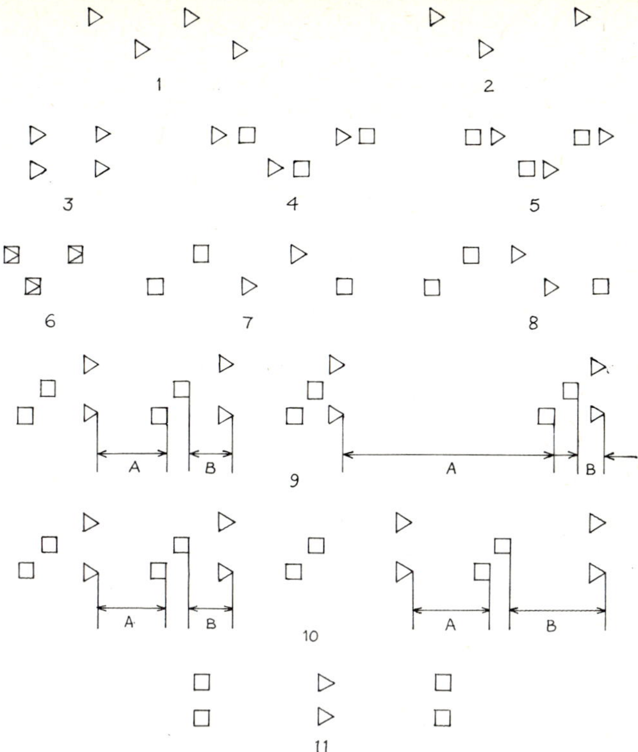

Fig. 9.2. Diagram of footfalls with different mammalian gaits. 1. Symmetrical walk and run; 2. half-paired and alternate ricochet; 3. paired ricochet; 4–6. symmetrical gaits; 7. diagonal gallop; 8. lateral gallop; 9. primitive ricochet at low and high running speeds; 10. half-bound at low and high running speeds; 11. bound. The animal is moving from left to right. Triangles denote tracks of hind feet, squares tracks of forefeet. A extended; B crossed stage of flight. Most zoologists do not distinguish the primitive ricochet from the half bound. After P. P. Gambaryan (1972), *How Mammals Run*. U.S.S.R.

One woman in Brisbane, Australia, managed to do so when a possum came down her chimney one night. It left clear sooty footmarks in trails over her white carpet, but she was not pleased with them.

Outdoors, the easiest way to record tracks is to photograph them against the sun in early morning or late afternoon. At these times the prints have a shadow which helps to accentuate them. A ruler should be placed beside one print so their measurements can be calculated from the photograph. Such measurements are not entirely accurate if a series of prints is being analysed, because the camera will show the central prints larger than those at the edges.

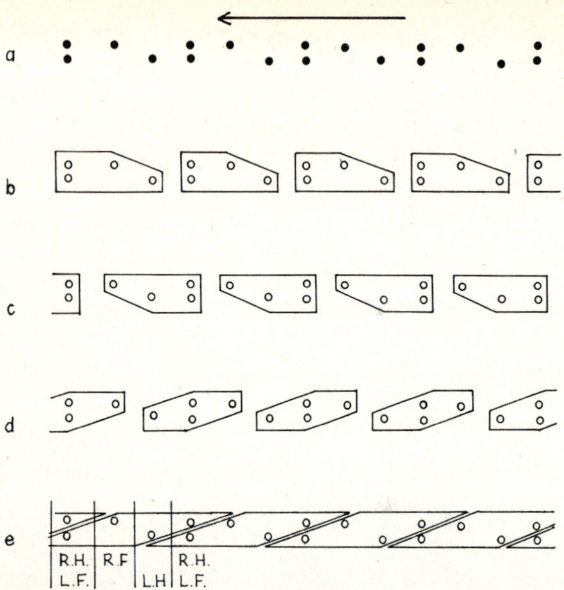

Fig. 9.3. Track of a horse and interpretations that can be made from it. (*a*) Track; (*b*)–(*d*) Erroneous interpretations; (*e*) Correct interpretation. Based on the work of E. Hékimian (1970), *Mammalia*, **34**, 118–135.

The best method of recording tracks is to do so directly, using a metre-rule. The six measurements that should be taken for each set of tracks, comprising five successive footprints, are given in fig. 9.4. With these should be included pertinent information on such things as the terrain, the substratum, the temperature, and if possible the size of the animal and its state of health. If the individual can be handled, its weight should be taken, together with measurements of its shoulder height, hip height, width at the shoulders, width at the hips, and trunk length. So far few tracks and the animals that made them have been accurately measured in this way, but when they have been, valuable comparisons about locomotion will be possible between species, and among individuals of one species. It will be interesting to determine what differences the breed of a species and its sex or state of health have on its tracks.

9.4. *Research Involving Tracks*

Zoologists who study small mammals have invented an ingenious way of studying the movements of an individual. Because schemes such as attaching radioactive material to a mouse and then tracking its position with a Geiger counter are cumbersome, and perhaps bad for the mouse, researchers now often clip one or more of a mouse's toes after it has been trapped in the wild. When it is released again, smoked paper can be set out in the vicinity of the release site, either along runways, or at feeding stations where food is left out. Every time the marked individual runs over this paper, its distinctive footprint is recorded so that the researchers know where it was on any one night. Using different patterns of clipping, many individuals can be studied at once, for information not only on home range, but on populations, association of individuals, association of species, longevity, and mortality.

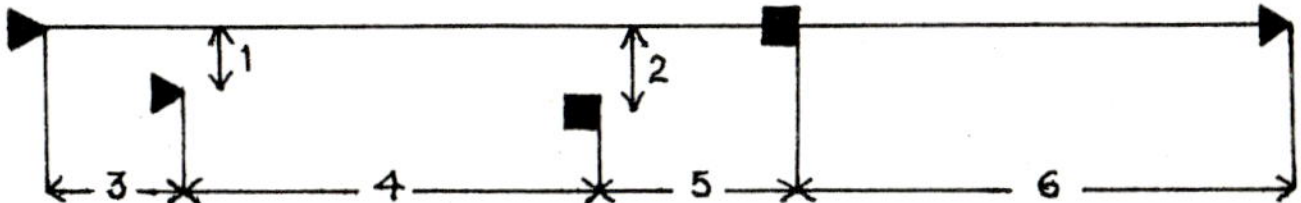

Fig. 9.4. Measurements to be taken for a sequence of footfalls. 1. Width of base of hind feet; 2. of forefeet; 3. distance between hind feet; 4. extended suspension; 5. distance between front feet; 6. flexed suspension. Triangles represent hind feet, squares front feet. After P. P. Gambaryan (1972), *How Mammals Run*. U.S.S.R.

Another example of research involving tracks could be carried out on sea turtles, without ever disturbing these animals. For example on Heron Island off Queensland in Australia both loggerhead and green turtles come ashore each night during the summer to lay eggs. By following their distinctive trails one could easily note where the eggs were laid, what species laid them, and how large the female was from the width of the track. Because the turtles come ashore only at fairly high tides, one could also work out at what time the turtle returned to the water (i.e. where the returning track ended because the turtle had become launched at that point) and often at what time it had left the water as well, if it had done so after high tide.

10. Invertebrate Locomotion

10.1. *Onychophoran Gaits*

Although arthropods possibly evolved from annelidan ancestors, their modes of locomotion bear little resemblance to those of annelids. Instead of relying on legs, leeches, earthworms and polychaetes tend to move by undulations useful in subterranean or aquatic environments. Some polychaetes have lateral parapodia which look superficially like useful legs, but these structures do not bend or retract readily and so cannot function well in walking or running, movements which demand that a leg shorten during each swing forward so that it does not strike the ground. While we may guess how the locomotion of the arthropods evolved from that of the annelids, the Phylum Onychophora supplies us with a form in many ways intermediate between the two phyla, and representative of the primitive arthropod stock from which has since evolved the various groups of living arthropods. The locomotion of *Peripatus*, a tropical and sub-tropical creature 50 mm long, resembling an active millipede, shows an advance on that of annelids, but it also displays great flexibility and undeveloped potentialities which are exploited by the more advanced Arthropoda. *Peripatus* progresses without any lateral undulations of the body, so no energy is wasted in side-to-side movement. The propulsive force is supplied by the extrinsic muscles to the legs which are under the body rather than lateral to it, and not by the longitudinal body musculature as in polychaetes. The length of the limbs is controlled by their intrinsic muscles and varies according to the phase of the step and gait being used. The length of the body can increase up to 12 per cent with increased speed of movement, but it is held fairly rigid by the musculature of the body wall.

S. M. Manton, of Cambridge and King's College London, studied the movements of *Peripatus* and *Peripatopsis* by using cinematography and recording their footprints. She found that these species have three main types of gait (fig. 10.1):

(*a*) *Bottom gear*, used to begin moving, in which many of the 19 or so pairs of legs are on the ground at one time, each for a relatively long

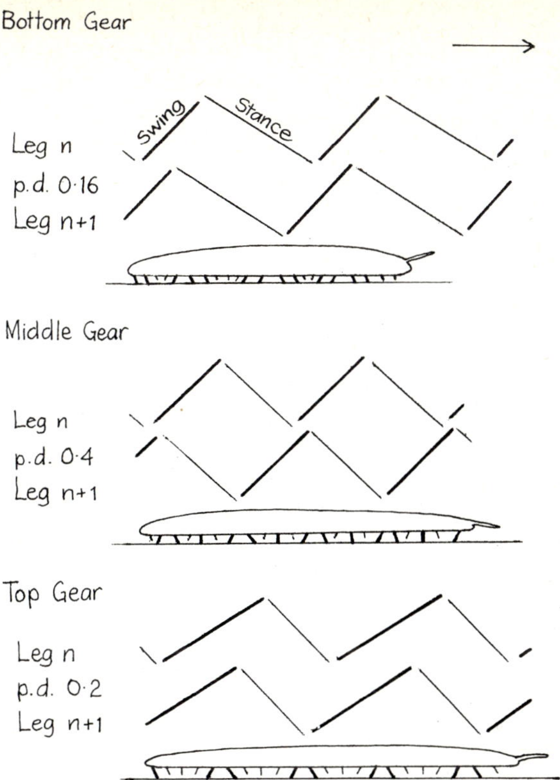

Fig. 10.1. Three gaits of *Peripatopsis*. Strokes of adjacent legs (*n*, *n*+1) and the phase difference (p.d.) between them are indicated. The legs which are on the ground are moving backward with respect to the body; the legs off the ground are swinging forward. Based on the work of S. M. Manton (1953), *Symposium of the Society of Experimental Biology*, **7**, 339–376.

period. Thus maximum power is obtained. This gait is analogous with that of a horse pulling a heavy wagon, in which each leg is set down for a long period as the animal strains forward.

(*b*) *Middle gear* produces speeds of 7–9 mm s^{-1}. The propulsive phase and swing phase of each leg are about equal, so that half the legs are on the ground at any one time. Each leg can swing to its maximum angle to increase its step. As in most animals each leg is put down just after the one before it is raised, so that the apparent wave of movement is from back to front. There is a large phase difference between successive legs.

(*c*) *Top gear* has fewer legs on the ground at any one time propelling the animal forward, so each leg has a longer period of swing than of propulsion. Thus each leg has more time to rest and touches the ground as little as possible so that friction will not retard the animal's forward speed. In this gait the phase difference between successive legs is adjusted so that each leg copies the movement of the one ahead, only fractionally later, so there is little chance of a collision between legs. In this gear paired legs are out of phase instead of moving in pairs as they do in bottom gear. It is easy for maintaining fast walking (about 10 mm s^{-1}) once this has been achieved.

10.2. *Primitive Arthropods*

In Arthropoda the general guiding principle behind the evolution of locomotion seems to be an increase of·speed. However, because of the exoskeleton, alterations in body extensions are impossible and changes in gaits are restricted. Various arthropod groups have taken up each of the three gaits of *Peripatus* and elaborated them; morphological adaptation in the course of evolution has often precluded a change in a gait once it has been taken up. The millipedes, which have many short body segments, have elaborated the low-gear gait of *Peripatus*; their movements are slow and power is needed in burrowing through ground litter. Some centipedes with relatively long segments and limbs have adopted the middle-gear gait; their movements are fast and agile. Other chilopods with quick movements have elaborated the top gear, in which the propulsive leg strokes are shorter than the swing strokes.

A jointed leg with a narrow tip, such as is present in many arthropods, is a type of limb superior to that of *Peripatus*, because a leg can be put down on almost the place occupied by the preceding leg, but before the preceding leg is raised. The weight of the body is thus immediately transferred to the leg behind, and is not briefly borne by more remote legs as in *Peripatus*. Most of the less specialized groups of short-legged arthropods such as the epimorphic Chilopoda show this advantage, which gives enough stability so that seven successive legs may be off the ground at once. The tendency of the body to sag between these points of contact must of course be countered by the trunk muscles.

10.3. *Insect Anatomy*

Insects represent the final stage in a process of limb reduction in the arthropods, with three pairs of thoracic legs. Because a minimum number of legs is best, as then the total weight of the limbs is least, one may wonder why the highly evolved insects do not have only two pairs of legs, as vertebrates do. It may be that two legs would not provide enough support for an insect, since they arise from the thorax rather than from both ends of the animal. With the decrease in leg numbers during insect evolution from many to six has come an increase in leg-length, which increases the length of a stride and thus the speed. However, a longer leg does not usually mean that an insect is taller, which might make it too vulnerable to air currents; instead, its body usually hangs down between the legs, so that the centre of gravity remains low and stable (fig. 10.2).

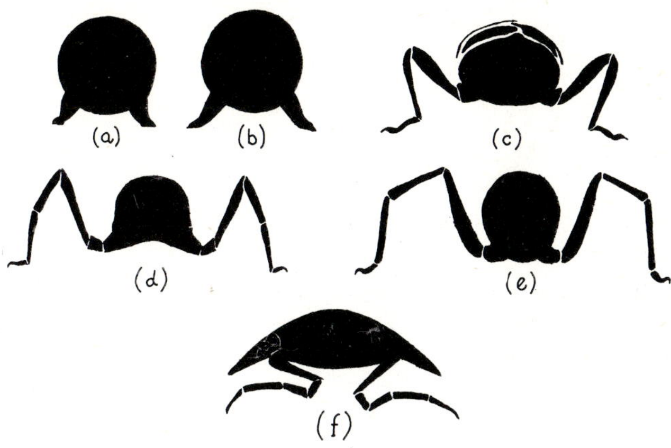

Fig. 10.2. Stances of some invertebrates depicted in cross-section. (*a*) Peripatus moving slowly; (*b*) Peripatus moving quickly; (*c*) Earwig; (*d*) Scorpion; (*e*) Spider; (*f*) Isopod. All of these animals have their bodies close to the ground, which makes them resistant to air currents. Based on the work of S. M. Manton (1953), *Symposium of the Society of Experimental Biology*, **7**, 339–376.

Legs can work effectively only if they have a rigid support to act against. In larval insects with soft bodies, leg muscles work against a body held firm by the turgor pressure of the haemolymph. Caterpillars, because of their length, have extra appendages or prolegs to help support them. Adult insects however have a chitinous exoskeleton which provides a strong but light support and to which the leg muscles are attached. As with most tetrapods, the proximal segments of each leg tend to be broad, with the distal end tapering towards the tip. Each leg is divided into four main segments. The most proximal, the coxa, usually articulates with the thorax by a dicondylic joint (moving in a single plane only), as does this segment with the adjacent tro-chanter–femur segment, and that with the tibia (fig. 10.3). However, the two former joints operate at right angles to each other, so that this part of the leg can move in many directions. The distal segment of the leg (the tarsus) is light and easily moved, as is this segment in all cursorial animals. The tarsus joins the tibia with a monocondylic joint, the nearest approach to a ball-and-socket joint found in invertebrates. Thus the insect leg has one more main segment than the basic vertebrate pentadactyl limb, which allows for greater extension. Also, whereas the mammalian legs are typically positioned under the trunk, those of insects are lateral to it.

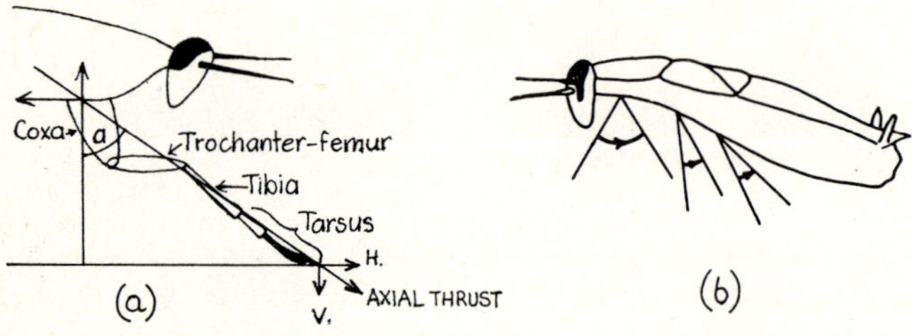

Fig. 10.3. Side views of the cockroach. (*a*) Diagram illustrating the forces acting on the tarsi when the foreleg functions as an inclined strut; $H =$ horizontal and $V =$ vertical forces. (*b*) Diagram indicating the range of angular movements of the three coxae during normal walking. From G. M. Hughes and P. J. Mill (1974), in *Physiology of Insecta*, Vol. 3, edited by M. Rockstein. London: Academic Press.

The musculature of insects differs from that of vertebrates because it is internal to the skeleton, not external. All the muscles, which are cross-striated and may pass over more than one joint, have a wide range in their rate of contraction, with both slow and fast motor fibres so that locomotion can occur over a continuous range of speeds. The activation of the muscles originates in the central nervous system, but as in vertebrates sensory feed-back from leg receptors is important for adjusting the basic patterns of activity.

As in all animals with legs, the legs function both as *struts* to support the body, and as *levers* to propel the body forward. These two components are difficult to separate, although in general there is a progressive increase in the lever action and a decrease in the strut action from anterior to posterior legs. The force of the tarsi on the ground can be divided into horizontal and vertical components, forces so minute that little work has yet been done on measuring them directly. The movements of the legs of a given insect are a function of the structure of each leg, especially of the joint between the thorax and the coxa. In the cockroach, an invaluable subject in research if not in other spheres of civilization, the first legs can move through much wider angles than the second legs, which in turn have a greater angular movement than the third pair of legs (fig. 10.3). The first coxa moves about an arc of 70° in the vertical plane, and the tarsi are placed on the ground well in front of the articulation with the thorax, so that the first legs have a horizontal component retarding progression. The second or meso-thoracic legs seldom and the third or metathoracic legs never have such a horizontal component during normal walking. A top view of a cock-roach indicates that the tarsi of the first legs are normally more or less oriented at 45° in a forward direction, the third legs at 45° in a backward direction, and the middle legs at right angles to the main axis of the cockroach body. The angles of the first and third legs are decreased in an insect walking on a slippery surface, just as the steps of a person walking on ice are smaller than usual. The third legs, because they are usually the longest in insects, must undergo the greatest change in length during walking or running. During extension they exert an axial thrust against the body and ground which propels the body forward.

10.4. *Insect Locomotion*

Because the limbs of insects are few and crowded together on the thorax, the original onychophoran gait has been lost and new gaits have appeared. It is impossible to distinguish between walking and running in insects: there is no change in the type of gait used at slow and at fast speeds, just a difference in the frequency at which the legs are moved and to a much lesser extent in the length of the strides. At its fastest speeds individual legs of a cockroach have a frequency of movement of 20 cycles per second (20 hertz, Hz).

Most insects, at least of the few species (from over a million) that have been studied, use all six legs when walking, the first largely for traction, the second for support, and the third for propulsion. Some however, such as some grasshoppers whose huge hind legs are specialized for jumping, use only the first four, while others, such as the preying mantis, whose front legs are specialized for obtaining food, use only the last four. All however walk by lifting the legs in an orderly sequence which remains fairly constant at a given speed. In general, the rhythm tends to be "an alternation of triangles of support", with the first and third legs of one side and the middle leg of the other supporting the insect in turn (fig. 10.4(*a*)). One can easily watch a small grasshopper or cricket moving this way. This gait is also used by the cockroach *Periplaneta americana* at all speeds from 50–800 mm s^{-1}. However, motion picture frames show that this pattern is not absolute by any means, because four and sometimes five legs may at one time also support a walking insect. If insects such as the cockroaches *Blatta* and *Periplaneta* and the dragonfly are walking very slowly, the legs of one side move first, the posterior legs moving before the legs ahead of them, as they do in vertebrates (fig. 10.4(*b*)). The movement of the legs in these and the

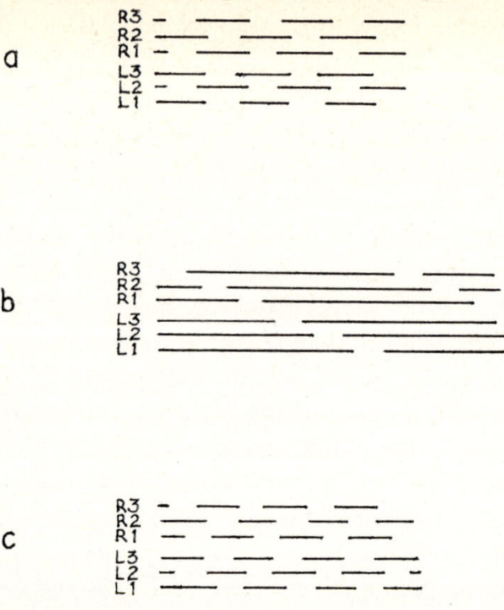

Fig. 10.4. Support diagrams of three gaits of insects. (*a*) Alternating tripod gait; (*b*) Slow gait with each leg stepping in turn; (*c*) Fast gait in which second stepping sequence is started before previous one has finished. R—right; L—left. Based on the work of G. M. Hughes and P. J. Mill (1974), in *Physiology of Insecta*, Vol. 3, edited by M. Rockstein. London: Academic Press.

many other sequences analysed in a variety of insects enabled G. M. Hughes and D. M. Wilson to formulate five rules which the legs followed:

(*a*) No first or second leg is lifted (protracted) until the leg behind has taken up its supporting position. Thus there is a posterior to anterior wave of moving legs on each side.

(*b*) Each leg alternates with its pair on the other side, so that they are half out of phase.

(*c*) The protraction or moving forward of a leg (= swinging time in vertebrates) is constant at all stepping frequencies.

(*d*) The retraction or supporting time of a leg decreases with an increase in stepping frequency.

(*e*) The stepping interval between third and second legs, and between second and first legs on each side remains relatively constant at all frequencies, whereas the interval between first and last legs decreases with an increase in frequency (fig. 10.4(*c*)). Such patterns are more or less innate, because insects placed on their backs show leg movements which are rhythmic and not random. They do not move the insect forward smoothly however. If the instantaneous speeds of an animal are calculated from high-speed films, it is evident that there is a rapid acceleration of the body in an alternating tripod gait after one set of feet touch the ground, while during the second half of the stroke the body quickly decelerates. Nor does an insect move in a perfectly straight line; rather when two legs are on the ground on the left side and one on the right, the animal veers slightly to the right, and *vice versa*, just as three people paddling in a canoe tend to steer the boat off course. Thus the centre of gravity of an insect swings slightly from side to side, just as it does in all walking or running vertebrates.

If two opposite legs of an insect are amputated, it tends quickly to assume the common tetrapod walking gait (left hind, left front, right hind, right front) which is also used in insects which habitually move with four legs. If only one leg is removed however there is little change in the animal's stepping pattern although the legs' postures may change to give the animal greater stability.

Terrestrial insects move much more slowly than one may imagine, perhaps because they can accelerate rapidly over short distances and change directions quickly. The top speed of a cockroach is only about five kilometres an hour. The speed is strongly dependent on temperature; H. Shapley found such a close relationship between temperature and speed of ants running in California that he could estimate to within 1°C the temperature from a single measurement of ant speed. This relationship held true over a range of 30°C. The speed also depends on the ground surface. Adult insects tend to reduce their speed when placed on glass although their first instar nymphs may double their speed on such a surface. This is because the nymphs rely on the adhesiveness of their tarsal pulvilli when walking, instead of on spines and claws used by adults.

Although all insect locomotion is generally defined as symmetrical, the left and right sides can operate independently, or become uncoupled, as they must when an insect turns. D. Graham of the University of Glasgow has studied this experimentally. By fastening a stick insect (*Carausius morosus*) with a stick glued to its back, he was able to watch its leg movements when it was lowered on to two treadmills. The right and left legs could move at different speeds if the treadmills which these legs touched were also going at different speeds, or one set of legs could be made to move while the other was stationary.

10.5. *Locomotion of Crabs*

The crab has been extensively studied because it is unique in running sideways, not tentatively, but very fast. Electromyographs made of the leg muscles in running crabs show there is asymmetry in muscle use: extensors and flexors in the meropodites of legs on the leading side frequently only maintain tonus, while those on the trailing side contract more actively, supplying virtually all the power for the movement. Rapid movement requires a large energy consumption, with legs cycling up to 20 Hz. Electrocardiograms with electrodes implanted through the carapace of ghost crabs (*Ocypode ceratophthalme*) showed that whereas the heart-rate of resting crabs ranged from 60–120 beats per minute, that of a running crab may reach 720 beats per minute. With speeds of 4 m s^{-1} (nearly 10 m.p.h.) ghost crabs are the fastest crustaceans on land, so fast that, as their name suggests, their moving shadow tends to be more noticeable than their sand-coloured body.

The gait of the ghost crab is not completely different from that of insects and spiders, even though the crab rushes sideways instead of forwards (fig. 10.5). The posterior legs move before those anterior and

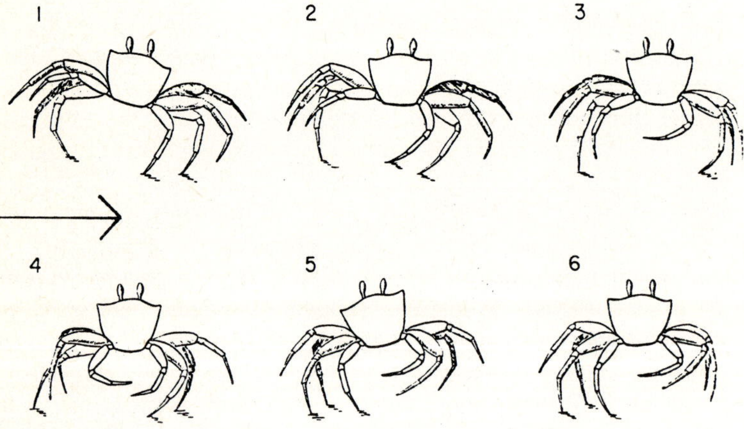

Fig. 10.5. Running of the ghost crab. The body tilts sideways as it is supported first by two legs on one side, and then three on the other. From D. R. Hafemann and J. I. Hubbard (1969), *Journal of Experimental Zoology*, **170**, 25–31.

adjacent to them, and legs adjacent to each other on one side of the body tend to move out of phase, so that the longest and largest legs, 2 and 3, do not usually both support the crab's body at any one time. The alternate pattern of support can be analysed for a crab by studying the marks left when the tip of each dactylopodite has been dipped in different colours of paint and the crab has run across a piece of white paper. Two legs might be expected to produce more thrust or pull than one leg, leading to long steps alternating with shorter ones, but successive steps were all about the same length. This may be because the single legs, 2 or 3, are the strongest. At fairly fast speeds the last two legs of the trailing side are commonly raised, so that only two legs, 2 and 3, are contributing thrust to the run and these are alternating. At highest speeds, amazing as it may seem, a ghost crab runs bipedally on only two legs, the trailing leg 3 supplying all the force while the leading leg 3 is used only for balance and support (fig. 10.6). During this gait often all the leading legs are off the ground at once and at intervals there are no legs on the ground at all, so the crab is in effect leaping along. The important difference between the running of insects and of crabs, besides the crab's leap, is that the legs bend in different ways so insects can go forwards and crabs sideways.

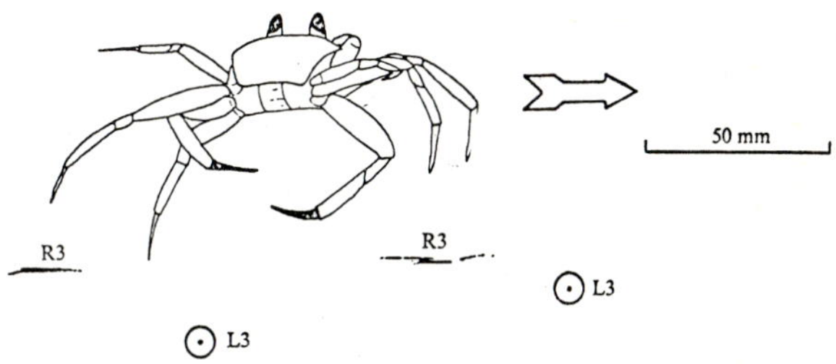

Fig. 10.6. Very fast running in the ghost crab, with tracks. The tip of the trailing leg (L3) which supplied the force leaves a sharp impression (circled), while that of the leading leg (R3) leaves a drag mark indicating it is used as a skid for balance without contributing sideways force to the movement. From M. Burrows and G. Hoyle (1973), *Journal of Experimental Biology*, **58**, 327–349.

Anatomically, the ghost crab is designed for speed because of its relatively long legs. M. Burrows and G. Hoyle noted in their research that the fastest ghost crabs were those with a carapace of about 20 mm across; crabs smaller than this ran more slowly because of their shorter legs, while larger crabs did so perhaps because of their heavier weight: a crab with a carapace 30 mm wide had legs 60 per cent larger but a body weight 400 per cent greater than one with a carapace 20 mm wide. However, the largest crabs were always males which, even when running, held their relatively larger claws in what was assumed to be an aggressive posture and frequently paused to rear up defensively. It may be that rapid running as an escape reaction in crabs triggered by any startle stimulus is replaced in part in large males by aggressive behaviour.

10.6. *Locomotion of a Crayfish*

Another invertebrate which walks with four pairs of legs and is as much at home in the water as on land is the crayfish. The British crayfish *Astacus* (*Austropotamobius*) *pallipes* can walk forwards, backwards, and sideways in both media. In this species the legs move through a small horizontal arc and the body is slung low between the legs, characteristics which help the animal to withstand the strong water currents sometimes encountered in streams. When they were walking under water, C. Pond of Oxford found that the crayfish moved forward smoothly, propelled by small thrusts by most or all of the walking legs in turn. On the land, where an individual's weight is 400–600 per cent greater than it is in the water, the animal lurched jerkily along, using any pair of legs which could obtain a grip to provide thrust. Since the mechanics of walking are very different on land and under water, one must be careful about generalizations for this gait in crayfish.

10.7. *Locomotion of a Spider*

If we compare the locomotion of invertebrates with eight legs with that of those with six, we find similar patterns. In spiders as in insects the stepping sequence passes toward the front along one side of the body, although if the posterior legs move again before a wave of movement has been completed by the anterior legs this sequence is obscured. In tarantulas (*Dugesiella hentzi?*), D. Wilson of California found there was much variation in walking, with a total of six possible stepping patterns in a stride. Although these spiders tend to have bilaterally opposite legs and segmentally adjacent legs used out of phase, there is much variation which is not, by and large, correlated with speed. When one or two of its legs were cut off or tied up, the tarantula changed the phase-coupling between the other legs to adapt to these indignities.

10.8. *Jumping in Invertebrates*

In the invertebrates, many unlikely species have incorporated jumping into their life style, a number of which will be discussed briefly here (fig. 10.7). Among insects, the best known jumpers are grasshoppers and locusts, fleas, click beetles, bristletails and springtails. When a grasshopper jumps, it raises the front part of its body with the first and second pairs of legs, then suddenly extends its enlarged hind legs by contracting the muscles in the femora which extend the femoral–tibial joint. Since these legs are long, they remain in touch with the ground and are therefore able to maintain a propulsive force for a relatively long time before the body is finally launched into the air. The herringbone arrangement of the leg muscles makes them very efficient; the measured take-off velocity of an adult locust weighing 3 g was $3 \cdot 40$ m s^{-1}, so that the force at the centre of gravity of the insect was $0 \cdot 43$ N and the thrust nearly seventeen times as great as the grasshopper's weight. The vertical and horizontal components of the take-off velocity determine the height and length of the jump. It is usually high, with a vertical component at up to 90°, because it frequently precedes flight where height is paramount. After high jumps a locust may land awkwardly, and sometimes on its back.

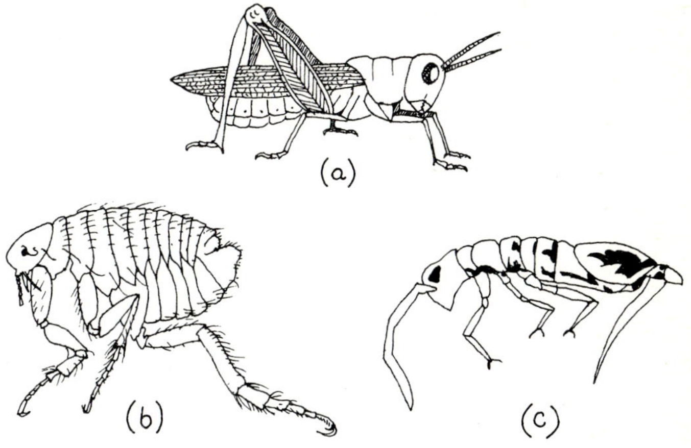

Fig. 10.7. Invertebrates that jump. (a) Grasshopper; (b) Flea; (c) Springtail.

Fleas also have enlarged hind legs for jumping, but these insects are so much smaller and more subject to air resistance than grasshoppers that direct muscular contractions alone cannot account for their jumping prowess, which can carry them 130 times their own height into the air. H. C. Bennet-Clark and E. C. A. Lucey of the University of Edinburgh found that fleas must invoke an energy-storage mechanism, using a resilin pad located on their body. This stored energy is released by a secondary muscle which then acts on the primary muscle which straightens each jumping leg. The rabbit flea, *Spilopsyllus cuniculus*, which needed $2 \cdot 5 \times 10^{-7}$ J of energy to jump a height of 35 mm, was capable of storing 4×10^{-7} J in each resilin pad. Larger species of fleas which jump higher than the rabbit flea have larger resilin pads and therefore more stored energy. Many authors have marvelled at the jumping ability of fleas, noting that if one were as big as a man it could jump over St. Paul's cathedral. In fact, however, the flea is only as proficient as it is because of its small size. If it were as big as a man, it would not be able to stand, let alone jump, because its relatively slender legs would collapse under the weight of its massive body.

The jump of the common click beetle or skipjack, *Athous haemor-rhoidalis*, is unique in that it is accomplished from the beetle's back, without the use of its legs. M. Evans of the University of Manchester, who studied this and other jumps of invertebrates in detail, found that it involved a jack-knifing movement which could spring the beetle 300 mm almost straight up. Before the jump the beetle bent its head and prothorax backward so that only the edges of its body touched the ground. Then, with a sudden jerk and click, it straightened its body out, the movement throwing the animal head over tail into the air. If it landed right side up, it walked away, but if it landed on its back, it repeated the jump. This jump is somewhat inefficient, as only 50–60 per cent of the energy expended by the beetle is used in raising it off the ground.

Sea-shore bristletails (*Petrobius brevistylis* and *P. marinus*) also jump with their bodies rather than their legs, but the thrust comes from the ventral surface of the abdomen pushing sharply against the ground. These jumps are used as they are in springtails, in escaping from predators. If bristletails are disturbed from under rocks where they pass the daytime, they usually perform a bewildering series of jumps, some more than 100 mm high, either forwards or backwards or sideways.

Springtails (Order Collembola) have evolved yet another way to jump. They have a forked structure or furcula on the underside of the fourth abdominal segment which at rest is folded forward under the abdomen where it is held in place by a clasp-like structure on the third abdominal segment called the tenaculum. To jump, a springtail suddenly extends the furcula ventrally and posteriorly. Individuals 5 or 6 mm long can jump 100 mm using this mechanism.

Finally, jumping fly larvae should be mentioned. The last instar of *Piophila casei*, which is called a jumper and lives in cheese, can leap as high as 200 mm into the air. The larva bends its head down beneath its abdomen and fastens its mandibles into its body near the posterior spiracles. The longitudinal muscles on the outside of this loop then contract, building up tension. When the mandibles release their hold suddenly, the larva jerks straight and is thrown upward.

The invertebrates mentioned so far in this section use their great jumping ability to escape from danger. Jumping spiders (*Salticidae*), by contrast, jump to capture food. They stalk up slowly on a small invertebrate and then leap on it, killing it with their fangs. Such a hunting method requires sure footing, so jumping spiders have pads of hair upon their feet, and secrete a safety-line which they fasten down at intervals as they move forward, as mountaineers do, to prevent a long fall. Unlike many jumping insects, jumping spiders lack any obvious specialization of their legs for jumping. Slow motion film by D. A. Parry and R. H. J. Brown of Cambridge of a male *Sitticus pubescens* jumping about 50 mm from a take-off to a landing platform

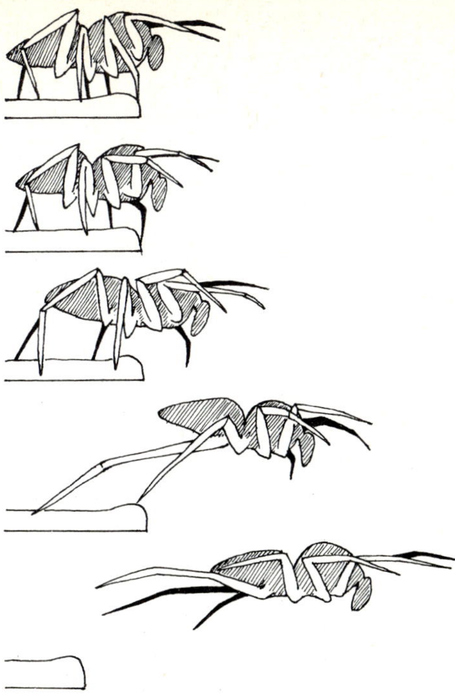

Fig. 10.8. Sequence of a male spider *Sitticus pubescens* jumping. Based on the work of D. A. Parry and R. H. J. Brown, *Journal of Experimental Biology*, **36**, 654–666.

show that its thrust comes mainly from its last pair of legs. These are held vertically before the jump, and straighten to their full length during take-off (fig. 10.8). The third legs, held lateral to the body, only help slightly to give the spider momentum, while the first and second legs are held off the ground and so do not contribute at all to the jump. The ending of the jump is variable. Because of tension in the safety drag-line of silk which the spider secretes to follow its flight, its path can be changed or its speed dampened. In one film sequence the thread seemed to tighten during flight, so that the jump ended abruptly with the spider dropping almost vertically. In another, in which the thread apparently broke, the spider ended the jump in a complete somersault.

Other spiders jump in other ways. The zebra spider *Salticus scenicus*, a common British species, uses the third legs more than the fourth, since the third pair leave the ground last, and other tropical spiders jump much farther, up to 250 mm, perhaps because they live in hot climates where invertebrate muscles can contract quickly.

From their films, Parry and Brown worked out the force of their spider's jump by measuring the spider's angle of take-off, the distance through which its body moved during take-off, and the velocity of take-off. From this they calculated the mean acceleration and, knowing the spider's mass, were able to work out the magnitude and direction of the force acting on the spider. This force was assumed to be evenly distributed between the feet of the fourth pair of legs as they suddenly straightened during take-off. From initial experiments with other spiders, it is known that there are no extensor muscles at the "hinge joints" (femur–patella and tibia–basitarsus joints) of the spider leg. Instead, extension in these joints is due to haemocoelic blood pressure in the leg and prosoma. Such a pressure system works at considerably less than optimum efficiency, and much less efficiently than do the muscles of such jumping insects as grasshoppers.

A millipede is perhaps the most unlikely jumping animal of all, but various tropical species of the sub-order Stemmiuloidea commonly employ a jump in their locomotion. When touched, one millipede runs forward, jumps 20–30 mm, lands, slides, and runs forward again. The jump is preceded by a rapid humping of the body about one-quarter of a body length behind the head. This hump becomes a loop which is thrown upward and forward, dragging the back half of the animal after it. The head and anterior leg segments maintain a grip on the ground until they are overtaken by the rapidly moving body loop, when they are also pulled up and forward. The animal lands in a U-shaped position and normally falls onto its side, but it quickly swings its head forward again preparatory to another run and jump. Perhaps the loop and jump evolved from the rapid spiralling movement of the body which the millipede may assume if it is threatened, with legs and head curled inside the body.

Further Reading

The following books on locomotion may be consulted:

Gambaryan, P. P. (1974). *How Mammals Run: Anatomical Adaptations*. New York: John Wiley. 367 pp. Translated from the Russian volume of 1972.

Gray, James (1953). *How Animals Move*. Cambridge: University Press. 114 pp. A short but useful introduction to the subject of all kinds of animal movements.

Gray, James (1968). *Animal Locomotion*. London: Weidenfeld and Nicolson, and New York: W. W. Norton. 479 pp. A comprehensive book including swimming and flying as well as terrestrial locomotion. Now unfortunately out of print.

Hildebrand, M. (1974). *Analysis of Vertebrate Structure*. New York: John Wiley. 710 pp.

Howell, A. B. (1944). *Speed in Animals*. New York and London: Hafner. 270 pp. Recently reprinted.

Sukhanov, V. B. (1974). *General System of Symmetrical Locomotion of Terrestrial Vertebrates and Some Features of Movement of Lower Tetrapods*. Washington: Smithsonian Institute; New Delhi: Amerind Publishing Co. Translated from the Russian volume of 1968.

Index

Scientific names are included when a specific animal has been discussed

Aardvark *Orycteropus afer* 67
Agouti *Dasyprocta* 69
Air resistance 102, 109, 123, 131
Albatross 91
Alexander, R. M. 21, 92, 111
Amble, see Running walk
Amphibian 16, 21–23, 25–26, 65, 81–84, 107–111
Anatomy, see Morphology
Annelida 120
Ant 127
Antelope 1, 65, 74, 99
   Saiga *Saiga tatarica* 14, 61, 99
Apes 20–21, 28–29, 68, 69–70, 77
Arboreal species 10, 14, 27–29, 45, 67–68, 74, 86, 91–92, 99, 101, 103
Argali, Pamir *Ovis ammon* 71
Aristotle 3
Arm movements 27, 28–29, 36, 43, 48–53, 55, 95
Armadillo *Dasypus novemcinctus* 24
   Giant *Priodontes giganteus* 67
Art, early 1–4
Arthropoda 120, 122–134
Artiodactyla 17, 63, 103
Ass, Mongolian wild *Equus hemionus* 11
Athletics 6, 48–55
Australopithecus 20–21, 29

Babbler *Picathartes* 92
Baboon *Papio* 14
Backbone, see Spinal column
Badger, European *Meles meles* 70
Badoux, D. M. 101–102
Bartholomew, G. A. 104–105

Basilisk *Basiliscus* 85–86
Bat 66
   Noctule *Nyctalus noctula* 74
   Vampire *Desmodus rotundus* 66–67
Bear 17, 69, 71
   Brown *Ursus arctos* 8–9
   Cave *Ursus spelaeus* 1
   Grizzly *Ursus arctos* 11
Behaviour 12, 14, 26, 99, 129
Bennet-Clark, H. C. 131
Bernshtein, N. A. 5
Biodynamics 31
Bipedalism 11–12, 23, 27, 77, 116–117, 129
   evolution of 28, 85–86
   in birds 77, 91–97
   in mammals 28–29, 67, 77, 99–107
   in reptiles 85–86, 89–90
   morphological adaptations 19–21, 99–100, 102–103, 104–107
Birds 1–2, 4, 11–12, 13, 16, 18, 23, 26, 27, 91–97, 116
Bison *Bison bonasus* 2
Blackbird *Turdus merula* 92
Bolen, E. G. 95–96
Bones 15–25
   strength of 21–23, 90, 112
Borelli, G. 4
Bound 5, 8, 16, 45, 67, 70, 71–75, 103, 105–107, 116–117
Bovidae 60–61
Brain 25–26, 77
Brachiation 28–29
Brandell, B. R. 48
Braune, W. 5
Bristletail *Petrobius* 130, 132

Brontosaurus 89
Brown, R. H. J. 132–134
Burrows, M. 129
Bush-baby *Galago* 99, 101

Cadence 33, 48
Calow, I. 21
Camel *Camelus dromedarius* 8,
13, 58, 61, 63, 66, 70–71, 113
Camelidae 1, 60–61, 71
Canidae 10, 75
Canter, see Gallop
Caribou *Rangifer tarandus* 10,
13, 61, 65, 66, 75, 77
Carnivora 18, 25, 103
Cary, G. R. 105
Cassowary *Casuarius* 91
Caswell, H. H. 104–105
Cattle 1, 116
Cavagna, G. A. 45, 47
Centipede 107, 122
Centre of gravity 31–33
displacement during walk 34–38,
40, 78–79, 87–88, 93–95, 127
displacement during fast gait 42,
44, 71, 74–75, 77, 79–80
displacement during jumping
51, 53–54, 98–103, 107–109,
111, 130
morphological 20, 27, 91, 123
Cervidae 60–61, 65, 66
Change of lead 73–74
Cheese jumper *Piophila casei* 132
Cheetah *Acinonyx jubatus* 5,
11–12, 14, 19, 74, 76–77, 116
Chelonia 87–89
Chilopoda 122
Chimpanzee *Pan* 29
Chinchilla *Chinchilla laniger* 26
Chinchillidae 67
Cineradiography 21–22
Clark, I. 92
Click beetle *Athous haemorrhoidalis*
130–132
Climbing 44–45, 74, 78, 101
Cockroach 124–125, 127
Blatta 125
Periplaneta americana 125
Compensation in gait 26, 40, 127,
130
Cormorant 91

Cottontail *Sylvilagus floridanus* 67
Cougar *Felis concolor* 115
Coyote *Canis latrans* 11
Crab, Ghost *Ocypode*
ceratophthalme 127–129
Crabbing 67
Crane 92
Crayfish *Astacus*
(Austropotamobius) pallipes
129–130
Cricket 125
Crocodilian 75, 81, 86–87
Crow *Corvus* 13
Cursorial species 10–11
birds 11–12, 77, 91
mammals 12–14, 16–19, 24–25,
76–77, 99
morphology 17–19, 91
reptiles 84–86
Cynognathus 90–91

Dagg, A. I. 61, 74, 79–80, 112
De Vos, A. 61
Deer 2, 5, 18, 62
Fallow *Dama dama* 61
Mule *Odocoileus hemionus*
11, 61, 78
Père David's *Elaphurus*
davidianus 61, 98
Red *Cervus elaphus* 61, 65
Sika *Cervus nippon* 61, 98
White-tailed *Odocoileus*
virginianus 11, 61
Digitigrade 17–18
Dingo *Canis familiaris* 104
Dinosaur 27, 85–86, 89–91
Display gaits 77–78
Distance running 49–50, 55
Dog *Canis familiaris* 18, 57–58,
62, 64, 67, 68, 71, 72, 76, 115
Alsatian or German shepherd
26, 63–64, 71, 111
Basset 64, 116
Beagle 68
Bloodhound 63–64, 71
Boxer 63–64
Collie 71
English bulldog 64
Great Dane 63–64, 71
Golden retriever 71
Greyhound 5, 11, 19, 72, 99, 111

Dog—*cont.*
  Rhodesian ridgeback 71
  Saluki 71
  Weimaraner 71
Dorcopsis 103
Dragonfly 125
Drawings 1–6, 48–50
Dromedary, see Camel
Duck 92
  Marine steamer *Tachyeres* 91
  Whistling *Dendrocygna* spp
    95–96

Earwig 123
Eberhart, H. D. 34
Echidna *Tachyglossus aculeatus*
    21–22
Eland *Taurotragus oryx* 75, 99
Electrogoniometer 30–31
Electromyography (EMG) 6–7,
    31–32, 39, 42–43, 127–128
Elephant 11, 57, 68, 70, 75
Elftman, H. 5
Elk ix, see Wapiti or Moose
Emu *Dromaius novae-hollandiae*
    91, 93
Endurance 1, 13–14, 69, 75, 105
Energy 19, 36–39, 75–77, 103, 107,
    131–132
  kinetic 38–39, 44–46, 51, 102,
    108–109
  potential 38–39, 44–46, 51
Errors in depiction of locomotion
    2–4, 5–6, 8–9, 115, 118
Euro *Macropus robustus* 104
Evans, M. 132
Evolution 81, 99
  of amphibians 110–111
  of birds 97
  of invertebrates 120, 122
  of mammals 12–14, 16–21,
    26, 27–29, 63, 67, 78, 103–107
  of reptiles 85–87, 89–91

Fenn, W. 48
Ferret *Mustela putorius* 21–22
Fischer, O. 5
Fish 110
  Bluegill sunfish *Lepomis* 83
Flea 130–131
  Rabbit *Spilopsyllus cuniculus*
    131

Flicker *Colaptes cafer* 92
Flight 4, 91, 95, 131
Footprints 17–18, 113–115,
    117–120, 128
Force 24–25, 45–47, 102, 108–109,
    124, 130–132
Force-plate or platform 23, 31,
    36–38, 40, 68, 92, 112
Fossorial species 24–25
Fowl, domestic 11–12
Fox 18, 113
  Gray *Urocyon cinereoargenteus*
    11
Freeman, M. A. R. 68
Frog 108
  African *Rana occipitalis* 83
  Bull *Rana catesbeiana* 109–110
  Common *Rana temporaria* 22
  Cricket *Acris crepitans* 83
  Green *Rana clamitans* 110
  Green tree *Hyla cinerea* 84
  South African *Rana oxyrhynchus*
    110
  Tree *Hyla crucifer* 110

Gait formula 62, 64, 73
Gallop 5–6, 13–14, 16, 19, 26,
    58, 59, 68, 70, 71–75, 76, 78,
    79–80, 81, 87, 114–117
  erroneous 6
  rotatory or lateral 71–72,
    116–117
  transverse or diagonal 71–72,
    116–117
Gambaryan, P. P. 32, 57, 71, 73,
    100, 111, 115–117
Gans, C. 109
Gazelle 19, 77, 78
  Grant's *Gazella granti* 61
  Thomson's *Gazella thomsoni* 11
Gecko 81
Geriatric gait 33, 41
Gericault, J. 5–6
Gibbon 28–29, 68
Giraffe *Giraffa camelopardalis*
    1–2, 10, 11, 13, 58, 60–62, 66,
    69, 70, 78–80, 93, 111–112, 116
Giraffidae 60–61, 63, 66
Goat *Capra hircus* 19, 26, 76
Goose, Magpie *Anseranas
    semipalmata* 93
Gorilla *Gorilla gorilla* 20–21, 29

Graham, D. 127
Grasshopper 108, 125, 130–131, 134
Graviportal locomotion 17, 68,
    70, 75
Greek athletics 48–53
Greenlaw, R. K. 8
Guillemot 91
Gull 92
  Silver *Larus novae-hollandiae*
    13, 93, 95

Habitat 10, 12–14, 28–29, 58,
    65–66, 71, 75, 78, 83, 86, 91, 97,
    98, 99, 100, 104, 106
Hafemann, D. R. 128
Half-bound 71, 74–5, 101, 116–117
Halteres 52–53
Hamster *Mesocricetus auratus*
    21–22
Hare 1, 5
  Arctic *Lepus arcticus* 101
  European *Lepus europaeus* 73
  Greenland *Lepus groenlandicus*
    101
Hartebeest *Alcelaphus* 69
Hékimian, E. 118
Heron, Reef *Egretta sacra* 92–93
Hildebrand, M. 19, 58, 62–64
Hippopotamus *Hippopotamus*
  amphibius 70, 75
Hobson, D. A. 8
Hop, see Jump
Horse *Equus caballus* ix, 2–3, 4,
    5–6, 7, 11, 17–18, 24, 62, 65, 68,
    69, 70, 71, 73, 76, 79, 111, 113,
    115, 116, 121
Hoyle, G. 129
Hubbard, J. I. 128
Hughes, G. M. 124, 126
Hummingbird 91
Hunting by man 1–2, 14, 29, 58, 115
Hunting dog, Cape *Lycaon pictus*
    11
Hurdle 51–52, 55, 112
Hutton, W. C. 68
Hyena 11, 12, 70, 75
  Spotted *Crocuta crocuta* 69
Hyrax, Tree *Heterohyrax brucei*
    21–22

Ibis, White *Threskiornis molucca*
    93–94

Impala *Aepyceros melampus* 14
Indris *Indri indri* 101
Inman, V. T. 34–39
Insect, anatomy 122–125
  larva 123
  locomotion 125–127, 128, 129
Invertebrate gaits 120–130
  jumps 130–134
Isopod 123

Jackal *Canis aureus* 11
Jackrabbit, Black-tailed *Lepus*
    californicus 67
  White-tailed *Lepus townsendii*
    67, 101
Jay *Garrulus glandarius* 92
  Blue *Cyanocitta cristata* 92
Jenkins, F. A. Jr. 21–22
Jerboa 77, 99, 100, 116
Jerboa marsupial *Antechinomys*
    spenceri 21, 106–107
Joint movement 19, 24–25, 30–31,
    40, 92–93
Jump 14, 17, 26, 27, 28, 46–47, 67,
    68, 77–78, 91, 92, 98–112
  High 54–55, 99, 105, 112
  Long 52–54, 55, 99, 102, 105,
    107, 109–110, 112
  of frog 21–23, 107–111
  of invertebrates 130–134
  of kangaroo 99–104

Kangaroo 1, 2, 8, 67, 68, 75,
    98–104, 108, 109, 116
  Black-faced *Macropus melanops*
    104
  Red *Megaleia rufus* 99
  Tree *Dendrolagus* 67, 75, 103
Kangaroo mouse *Microdipodops*
    100
Kangaroo rat *Dipodomys* 68,
    77, 99, 104–106
Kinematics 29–31
Kinetics 31
Kinkajou *Potos flavus* 67
Kiwi *Apteryx* 91
Kookaburra *Dacelo novaeguineae*
    92

Lagomorpha 67, 101
Landrail, Banded *Rallus*
    phillippensis 93–96

Lawrence, M. J.   67
Leap, see Jump
Lechwe   *Kobus leche*   61, 66, 78
Lemur, Ring-tailed   *Lemur catta*
   69
   Weasel   *Lepilemur*   101
   Woolly   *Avahi laniger*   101
Leonardo da Vinci   4
Leopard   *Panthera pardus*   14
Letts, R. M.   7, 30, 32, 42–43
Lion   *Panthera leo*   8, 11, 12, 14,
   78
Lizard   13, 27, 84–86, 87, 90
   Collared   *Crotaphytus collaris*
      84–86
   Teiid   *Ameiva ameiva*   86
Llama   *Lama glama*   71
Locust   130–131
Lope, see Gallop
Lucey, E. C. A.   131

Macropodidae   103–104
Magpie   *Gymnorhina tibicen*   94
Mammals   16, 21, 25–26, 56–80
Mammoth   1
Man, Primitive   1–2, 113
Man   *Homo sapiens*   4–5, 6–8,
      11–12, 16, 17, 18, 19–21, 27–55,
      77, 78, 86, 92–93, 113
Mantis, Praying   125
Manton, S. M.   120–121, 123
Marey, E. J.   4, 30
Margaria, R.   45, 47
Marlow, B. J.   106–107
Marsupialia   1, 27
Martin   91
Mech, D.   12
Metabolism   41, 76–77, 81, 128
Migration   10, 13–14
Mill, P. J.   124–126
Millipede   120, 122, 134
Moa   1
Monkey   28, 69, 77
   Capuchin   *Cebus capucinus*   77
   Spider   *Ateles*   77
Moorhen   *Gallinula chloropus*   93
Moose   *Alces alces*   ix, 12, 61, 66, 75
Morphology   15–25, 28–29, 63,
      69–71, 74–75, 86, 89, 102–103,
      105–106, 108–109, 110–111,
      122–125

Mouse   45, 76, 118–119
   Jumping (Zapodidae)   74
   Marsupial   74
Murray, M. P.   41
Muscle, Gross action   4, 15–16,
      20–22, 24–25, 31–32, 38–40,
      42–43, 49, 68
   Gross structure of   15, 19, 23
   Microscopic structure and activity
      6, 11, 15, 23, 31
Mustelidae   70, 74, 75
Muybridge, E.   5, 67
Mynah Bird   *Acridotheres tristis*
   92

Napier, J.   20, 28
Neck movements   78–80, 94–96,
      111–112
Nervous control   15, 25–26, 28, 92,
      124
Newcastle, P. G.   4
Newts   82
Nightjar   91
Nodding of head in birds   93–95
Notomys cervinus   106–107

Okapi   *Okapia johnstoni*   60–61,
      66, 69
Onychophora   120–122, 125
Opossum   *Didelphis marsupialis*
      21–22, 75
Orang utan   *Pongo*   29
Ostrich   *Struthio camelus*   1–2,
      11–12, 91, 93, 116
Oystercatcher, Black   *Haematopus
      fuliginosus*   93

Pace   3, 8, 13, 57, 58, 59, 62, 63,
      64, 70–71, 75, 76
Pangolin, African   *Manis*   68, 77
Parrot   91
Parry, D. A.   132–134
Pathological gaits   6–7, 13, 23, 26,
      31, 40–43, 98, 127, 130
Peafowl   *Pavo cristatus*   95
Pendulum motion   4, 19, 38, 41,
      68, 78, 87
Penguin   91, 92
   Rockhopper   *Eudyptes crestatus*
      92
Peripatopsis   120–121

Peripatus 120, 122–123
Perissodactyla 17, 103
Phalangeridae 103
Photography 4–5, 6–8, 29–30, 57–59, 65, 66, 81–83, 98, 104, 109, 117, 120, 132, 134
Physiology 25–26, 76–77
Pig *Sus scrofa* 11
Pigeon *Columba livia* 93
Pittas Pittidae 92
Plantigrade 17–18
Platypus *Ornithorhynchus anatinus* 75
Plover, Spur-winged *Lobibyx novae-hollandiae* 95
Pocket mouse *Perongathus* 21, 105–106
Polychaete 120
Pond, C. 130
Possum Phalangeridae 117
Posture of ducks 96–97
of invertebrates 123
of primates 28
of reptiles 82, 86
Prance 78
Prehistoric pictures 1–2
Primates 27–29, 66, 78
Principles of locomotion 10, 56–57
Proconsul 29
Pronghorn antelope *Antilocapra americana* 11, 14, 61, 65, 66, 78, 99
Pronk, see Stott
Prosimian 77, 99, 101

Quail *Coturnix coturnix* 92
Quanbury, A. O. 7, 30, 32, 42–43
Quokka *Setonix brachyurus* 75, 103, 116

Rabbit 11, 13, 25, 26, 67, 113, 116
European *Oryctolagus cuniculus* 68
Marsh *Sylvilagus palustris* 67
Snowshoe *Lepus americanus* 67
Swamp *Sylvilagus aquaticus* 67
Racial differences in man 55
Rack, see Pace
Rail Rallidae 91
Raptor 91, 92

Rat *Rattus norvegicus* 21–22, 23
Ratel *Mellivora capensis* 70
Raven *Corvux corax* 92
Reindeer *Rangifer tarandus* 11, 65
Reptiles 16, 26, 27, 65, 75, 81, 84–91
Rhea *Rhea americana* 77, 91, 93
Rhinoceros 1, 17, 70, 75
Ricochet 27, 83, 99–107, 116–117
Right vs left 3, 42, 56, 127
Robin, American *Turdus migratorius* 92
Rowntree, V. J. 76
Running, Amphibian 81
Bird 1–2, 91, 96
Crab 128–129
Human 16, 34, 42–45, 48–51
on moon of man 46–47
of quadrupeds, see Gallop
Reptile 81, 85
Running walk 68
Rylander, M. K. 95–96

Salamander 81, 110
Spotted *Salamandra salamandra* 82–84
Saltatorial locomotion, see Jumping
Saunders, J. B. D. M. 34
Savage, R. J. G. 24
Scent glands 78
Sciuridae 70, 74, 75
Scorpion 123
Seton, E. T. 113
Shapley, H. 127
Shearwaters 91
Sheep 1
Sherrill, C. 51
Shoulder girdle 15, 16, 19, 24, 67, 83
Sitatunga *Tragelaphus spekii* 98
Skeleton 16, 19–20, 28, 122–124
Skifaka *Propithecus* 101
Skittering on water 83–84, 86
Skunk, Striped *Mephitis mephitis* 75
Skylark *Alauda arvensis* 92
Sloth 1, 10, 68
Three-toed *Bradypus* 11
Smith, J. M. 24
Snyder, R. 84–86

Souslik, Siberian   *Citellus
      undulatus*   73
Sparrow, House   *Passer domesticus*
      92
Speed   56, 59, 63, 76, 110, 114, 129
   maximum   10–11, 86, 87, 91,
      104, 107, 125, 127, 128
   selection for   12–14, 85–86, 92
   within a gait   33, 44, 46, 59, 76,
      88, 93, 120–122
   within a stride   83, 88–89, 127
Spider   11, 107, 123, 128, 130, 133
   Jumping   *Sitticus pubescens*
      132–134
   Tarantula   *Dugesiella hentzi?*
      130
   Tegenaria atrica   11
   Zebra   *Salticus scenicus*   133
Spinal column   5, 16, 19–21, 25,
      67, 69, 70, 74–75, 82–83, 84, 87,
      110
Sports   55
Spring hare   *Pedetes capensis*   67,
      100
Springbok   *Antidorcas marsupialis*
      14
Springtail   130–132
Sprint   11, 48–49, 51–52, 55, 76–77
Spronk, see Stott
Squirrel   11, 13
Stance   9, 16–18, 21, 27, 84, 86, 87,
      89–91, 123–124
Start, crouch   51
Stemmiuloidea   134
Step   33, 48, 78
Stick insect   *Carausius morosus*   127
Stilt   92
   White-headed   *Himantopus
      himantopus*   93
Stott   77–78
Stress   16, 73, 112
Stride   33, 41, 49, 56, 61, 71, 86,
      89–90
Struthiomimus   89–90
Substrate   46, 51, 113–114
   marsh or mud   26, 67, 74, 75, 98,
      115
   rough   66, 75, 98, 107
   slippery   51, 125
   smooth   58, 71
   snow   26, 66, 74, 114–115

Sukhanov, V. B.   81–82
Swallows   91
Swanson, S. A.   68
Swift   91
Swimming   10, 26, 47, 55, 83, 87,
      89, 91, 97, 111, 119

Tail, amphibian   81–83
   bird   91, 93–94
   mammal   23, 68, 77, 78, 100–107
   reptile   81, 85–86, 87, 89
Tapir   *Tapirus*   17
Tarsier   *Tarsius*   28, 99, 101
Taxonomy   23, 88, 95–96, 103, 110
Taylor, C. R.   76
Taylor, M.   13
Tayra   *Tayra barbara*   70
Technology, modern   6–8, 21,
      29–31
Temperature   65, 76–77, 81, 127
Toad, Southern   *Bufo terrestris*
      110
Tortoise   1, 11, 62, 87–89
Tracks   89, 113–119, 128–129
Tree shrew   *Tupaia glis*   21–22
Trot   5, 8, 14, 26, 57–58, 59, 62–64,
      68, 69–70, 71, 75, 76, 81, 83,
      84–85, 88, 114
Tsessebe   *Damaliscus lunatus*   61
Turtle   87–89, 113
   Green   *Chelone mydas*   89, 119
   Loggerhead   *Caretta caretta*
      89, 119
   Painted   *Chrysemys*   88, 89
   Snapping   *Chelydra*   88
   Soft-shelled   *Trionyx*   88
Tyrannosaurus   89–90

Uccello, P.   4
Ungulates   3, 13, 17–18, 19, 25,
      59, 60, 75, 78, 103
Unguligrade   17–18
Ursidae   75

Vasari, G.   4
Vicuña   *Vicugna vicugna*   71
Viverridae   13, 69
Vole   *Microtus pennsylvanicus*   70

Walk   116–117
   abnormal   41, 43

amphibian 81–82
bird 91, 93–95
erroneous 2–4, 8–9
human 7, 8, 29–42, 44
on moon, of man 45–47
quadrupedal 3, 14, 56–68, 78, 79, 80, 114, 117
reptile 81, 88
Walk pattern 60–61, 63–66, 72
Walker, W. F. Jr. 88
Wallaby 67, 68, 75, 103
Wapiti *Cervus canadensis* ix, 11, 61, 75
Wart hog *Phacochoerus aethiopicus* 11
Waterbuck *Kobus* 75
Weber, W. and E. 4
Wildcat *Felis sylvestris* 115

Wildebeest *Gorgon taurinus* 10, 11, 12, 14, 61, 69, 73–74
Wilson, D. M. 126, 130
Windsor, D. E. 103
Winter, D. A. 7–8, 30, 32, 42–43
Wolf *Canis lupus* 12, 13, 14, 113
Wolverine *Gulo gulo* 70, 75
Woman, in high heels 40
pregnant 33–34
Woodchuck *Marmota monax* 75
Woodpecker *Picus viridis* 92
Work 38, 44–45

Yak *Poephagus grunniens* 98

Zebra 11
Zoo 58, 65, 69, 98, 103
Zug, G. 87, 110

THE WYKEHAM SCIENCE SERIES

1   *Elementary Science of Metals*   J. W. MARTIN and R. A. HULL
2   †*Neutron Physics*   G. E. BACON and G. R. NOAKES
3   †*Essentials of Meteorology*   D. H. McINTOSH, A. S. THOM and V. T. SAUNDERS
4   *Nuclear Fusion*   H. R. HULME and A. McB. COLLIEU
5   *Water Waves*   N. F. BARBER and G. GHEY
6   *Gravity and the Earth*   A. H. COOK and V. T. SAUNDERS
7   *Relativity and High Energy Physics*   W. G. V. ROSSER and R. K. McCULLOCH
8   *The Method of Science*   R. HARRÉ and D. G. F. EASTWOOD
9   †*Introduction to Polymer Science*   L. R. G. TRELOAR and W. F. ARCHENHOLD
10   †*The Stars; their structure and evolution*   R. J. TAYLER and A. S. EVEREST
11   *Superconductivity*   A. W. B. TAYLOR and G. R. NOAKES
12   *Neutrinos*   G. M. LEWIS and G. A. WHEATLEY
13   *Crystals and X-rays*   H. S. LIPSON and R. M. LEE
14   †*Biological Effects of Radiation*   J. E. COGGLE and G. R. NOAKES
15   *Units and Standards for Electromagnetism*   P. VIGOUREUX and R. A. R. TRICKER
16   *The Inert Gases: Model Systems for Science*   B. L. SMITH and J. P. WEBB
17   *Thin Films*   K. D. LEAVER, B. N. CHAPMAN and H. T. RICHARDS
18   *Elementary Experiments with Lasers*   G. WRIGHT and G. FOXCROFT
19   †*Production, Pollution, Protection*   W. B. YAPP and M. I. SMITH
20   *Solid State Electronic Devices*   D. V. MORGAN, M. J. HOWES and J. SUTCLIFFE
21   *Strong Materials*   J. W. MARTIN and R. A. HULL
22   †*Elementary Quantum Mechanics*   SIR NEVILL MOTT and M. BERRY
23   *The Origin of the Chemical Elements*   R. J. TAYLER and A. S. EVEREST
24   *The Physical Properties of Glass*   D. G. HOLLOWAY and D. A. TAWNEY
25   *Amphibians*   J. F. D. FRAZER and O. H. FRAZER
26   *The Senses of Animals*   E. T. BURTT and A. PRINGLE
27   †*Temperature Regulation*   S. A. RICHARDS and P. S. FIELDEN
28   †*Chemical Engineering in Practice*   G. NONHEBEL and M. BERRY
29   †*An Introduction to Electrochemical Science*   J. O'M. BOCKRIS, N. BONCIOCAT, F. GUTMANN and M. BERRY
30   *Vertebrate Hard Tissues*   L. B. HALSTEAD and R. HILL
31   †*The Astronomical Telescope*   B. V. BARLOW and A. S. EVEREST
32   *Computers in Biology*   J. A. NELDER and R. D. KIME
33   *Electron Microscopy and Analysis*   P. J. GOODHEW and L. E. CARTWRIGHT
34   *Introduction to Modern Microscopy*   H. N. SOUTHWORTH and R. A. HULL
35   *Real Solids and Radiation*   A. E. HUGHES, D. POOLEY and B. WOOLNOUGH
36   *The Aerospace Environment*   T. BEER and M. D. KUCHERAWY
37   *The Liquid Phase*   D. H. TREVENA and R. J. COOKE
38   †*From Single Cells to Plants*   E. THOMAS, M. R. DAVEY and J. I. WILLIAMS
39   *The Control of Technology*   D. ELLIOTT and R. ELLIOTT
40   *Cosmic Rays*   J. G. WILSON and G. E. PERRY
41   *Global Geology*   M. A. KHAN and B. MATTHEWS
42   †*Running, Walking and Jumping: The science of locomotion*   A. I. DAGG and A. JAMES
43   †*Geology of the Moon*   J. E. GUEST, R. GREELEY and E. HAY
44   †*The Mass Spectrometer*   J. R. MAJER and M. P. BERRY
45   †*The Structure of Planets*   G. H. A. COLE and W. G. WATTON
46   †*Images*   C. A. TAYLOR and G. E. FOXCROFT
47   †*The Covalent Bond*   H. S. PICKERING
48   †*Science with Pocket Calculators*   D. GREEN and J. LEWIS

THE WYKEHAM ENGINEERING AND TECHNOLOGY SERIES

1   *Frequency Conversion*   J. THOMSON, W. E. TURK and M. J. BEESLEY
2   *Electrical Measuring Instruments*   E. HANDSCOMBE
3   *Industrial Radiology Techniques*   R. HALMSHAW
4   *Understanding and Measuring Vibrations*   R. H. WALLACE
5   *Introduction to Tribology*   J. HALLING and W. E. W. SMITH

All orders and requests for inspection copies should be sent to the appropriate agents. A list of agents and their territories is given on the verso of the title page of this book.

†*(Paper and Cloth Editions available.)*